Titanium Dioxide Nanomaterials

MATERIALS RESEARCH SOCIETY
SYMPOSIUM PROCEEDINGS VOLUME 1352

Titanium Dioxide Nanomaterials

Symposium held April 25–29, 2011, San Francisco, California, U.S.A.

EDITORS

Xiaobo Chen

Lawrence Berkeley National Laboratory
University of California–Berkeley
Berkeley, California, U.S.A.

Michael Graetzel

Ecole Polytechnique Fédérale de Lausanne
Lausanne, Switzerland

Can Li

Chinese Academy of Sciences
Dalian Institute of Chemical Physics
Dalian, China

P. Davide Cozzoli

Università del Salento and National Nanotechnology
Laboratory-Nanoscience Institute of CNR
Lecce, Italy

Materials Research Society
Warrendale, Pennsylvania

CAMBRIDGE
UNIVERSITY PRESS

CAMBRIDGE
UNIVERSITY PRESS

Shaftesbury Road, Cambridge CB2 8EA, United Kingdom

One Liberty Plaza, 20th Floor, New York, NY 10006, USA

477 Williamstown Road, Port Melbourne, VIC 3207, Australia

314–321, 3rd Floor, Plot 3, Splendor Forum, Jasola District Centre, New Delhi – 110025, India

103 Penang Road, #05–06/07, Visioncrest Commercial, Singapore 238467

Cambridge University Press is part of Cambridge University Press & Assessment, a department of the University of Cambridge.

We share the University's mission to contribute to society through the pursuit of education, learning and research at the highest international levels of excellence.

www.cambridge.org
Information on this title: www.cambridge.org/9781605113296

Materials Research Society
506 Keystone Drive, Warrendale, PA 15086, USA
http://www.mrs.org

First published 2012

CODEN: MRSPDH

A catalogue record for this publication is available from the British Library

ISBN   978-1-605-11329-6   Hardback

CONTENTS

PROPERTIES

APPLICATIONS

*Invited Paper

*Invited Paper

PREFACE

Titanium dioxide nanomaterials have been showing very promising applications in many fields. Scientists from all over the world have gathered together to discuss the most recent advances in areas spanning from theoretical calculations, to fundamental surface science, and materials fabrication, characterization and practical exploitation. Over 160 papers, presentations and posters, were accepted in Symposium GG, "Titanium Dioxide Nanomaterials" at the 2011 MRS Spring Meeting held April 25-29 in San Francisco, California. Among the invited presentations, Professor Annabella Selloni from Princeton University has shown the importance of the TiO_2/water interface. Professor John Yates from the University of Virginia delivered an exciting talk on the surface science of TiO_2 and the related electronic excitation and deexcitation processes. Professor Ulrike Diebold from Institute for Applied Physics, Vienna, Austria and Tulane University introduced us to the fundamental aspects of the organic molecule adsorption and reaction on TiO_2 surfaces. Professor Kazunari Domen from the University of Tokyo and Professor Hiroaki Tada from Kinki University, Japan demonstrated how to apply TiO_2-based nanomaterials for generating hydrogen from water under sunlight irradiation. Dr. Hugo Destaillats from Lawrence Berkeley National Laboratory discussed the use of TiO_2 photocatalysts in indoor air cleaning applications and the related challenges and opportunities.

In this printed proceeding are some of the selected papers which cover the synthesis, properties, and applications of titanium dioxide nanomaterials.

Xiaobo Chen
Michael Graetzel
Can Li
P. Davide Cozzoli

September 2011

MATERIALS RESEARCH SOCIETY SYMPOSIUM PROCEEDINGS

MATERIALS RESEARCH SOCIETY SYMPOSIUM PROCEEDINGS

Prior Materials Research Society Symposium Proceedings available by contacting Materials Research Society

Calculation

Mater. Res. Soc. Symp. Proc. Vol. 1352 © 2011 Materials Research Society
DOI: 10.1557/opl.2011.1007

First-principles study of Oxygen deficiency in rutile Titanium Dioxide

Hsin-Yi Lee,[1] Stewart J. Clark,[2] John Robertson[1]

[1]Engineering Department, Cambridge University, Cambridge, CB2 1PZ, UK
[2]Physics Department, Durham University, Durham, DH1 3LE, UK

ABSTRACT

The energy levels of the different charge states of an oxygen vacancy and titanium interstitial in rutile TiO_2 were calculated using the screened exchange (sX) hybrid functional [1]. The sX method gives 3.1 eV for the band gap of rutile TiO_2, which is close to the experimental value. We report the defect formation energy of the oxygen deficient structure. It is found that the defect formation energies, for the neutral charge state, of oxygen vacancy and titanium interstitial are quite similar, 2.40 eV and 2.45 eV respectively, for an oxygen chemical potential of the O-poor condition. The similar size of these two calculated energies indicates that both are a cause of oxygen deficiency, as observed experimentally [2]. The transition energy level of oxygen vacancy lies within the band gap, corresponding to the electrons located at adjacent titanium sites. The sX method gives a correct description of the localization of defect charge densities, which is not the case for GGA [3-6].

INTRODUCTION

Titanium dioxide (TiO_2) is a widely used material ranging from a substance used in solar cells, photocatalysts and nano-scale electronic devices. It has received a great deal of attention because it possesses many useful properties, such as high dielectric constant, good chemical stability, low cost and high refractive index [7]. Understanding the electronic and structural properties of the bulk and defective structures of TiO_2 is essential to improve the practical applications.

There are three different polymorphs of TiO_2 in nature: rutile, anatase, and brookite. Out of the three phases, the rutile is the most abundant naturally occurring phase and is the stable state under atmospheric conditions [3], hence it has been chosen as the major subject of this work.

Figure 1 shows the crystal structure of rutile TiO_2.

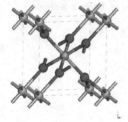

Figure 1. The unit cell of rutile TiO_2. Red spheres represent O atoms and light grey spheres represent Ti atoms.

First-principles calculations provide the methods that can be used to simulate a wide range of material properties. For many years, the local-density approximation (LDA) and generalized gradient approximations (GGA) within density functional theory (DFT) provided an efficient method for such calculations in solids, giving both lattice constants and bulk moduli with reasonable accuracy [1]. However, even though LDA and GGA gives acceptable crystal structures and ground state properties, it has the well-known band gap errors [8-11]. A more accurate description of the electronic structures is required. In this paper, we use a non-local exchange-correlation functional, the screened exchanged (sX) method [12], in order to correct the size of band gap and to improve the computational accuracy for TiO2.

METHOD

The sX method is based on the Hartree-Fock method for the exchange potential energy that is separated into two terms. One term is Thomas-Fermi screened-exchange potential, and the other is the remainder. The remainder and correlation potential energy can be calculated by the LDA, while the Thomas-Fermi screened-exchange potential can be evaluated from:

$$V_C^{sX-LDA}(r,r') = -\sum_i \frac{\phi_i(r)exp-(k_{TF}|r-r'|)\phi_j^*(r)}{|r-r'|} \quad (1)$$

In Equation (1), V_C^{sX-LDA} is the non-local screened-exchange potential energy, ϕ are the Kohn-Sham orbitals, and k_{TF} is the Thomas-Fermi screening wave vector. The exchange-correlation hole can be calculated in an accurate way by using the Thomas-Fermi screening of the Hatree-Fock exchange potential energy, so it is consequently become non-localized electrons system compare with the LDA. Moreover, sX modified the non-existing discontinuity errors of exchange- correlation potential from LDA. Thus, the sX method improves the simulation accuracy for the value of energy band gap and has been found to give good results for some semiconductors and insulators.

In this work, we constructed a 2x2x2 supercell of 48 atoms of rutile TiO$_2$, containing a single oxygen vacancy or titanium interstitial. The oxygen vacancy is created by taking out one oxygen atom from the supercell; and similarly, an extra titanium atom is placed into the supercell to form the titanium interstitial model. The defect structures are relaxed in their various charge states by sX functional using the standard norm-conserving pseudopotentials, and the energy levels and density of states are then obtained through the optimized structures. Cut-off energy 750 eV was used which was decided by the energy convergence test. For the k- point integration, we specified at Γ point. A Γ k-point selection indicates that the calculation is doing at a single k-point (0,0,0) in the Brillouin zone.

4

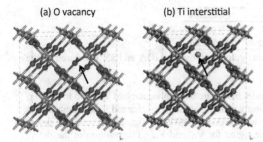

(a) O vacancy   (b) Ti interstitial

Figure 2. The crystal structures for rutile TiO_2 of (a) O vacancy, and (b) Ti interstitial. The defect sites are indicated by the arrows.

RESULTS

The band structure calculated by sX is shown in Figure 3, and Table 1 compares the calculated band gaps of rutile TiO_2 between sX, GGA and the experiment value. It can be seen that the band gap improves significantly from 1.86 eV in GGA to 3.1 eV in sX. The sX gap is close to experiment now. It is often thought that the band gap error is about 30%. This is true for Si, but for some oxides such as TCOs the error can be closer to 70-80% [13]. For rutile TiO_2, the gap opens up by 40% from GGA to sX in our calculations.

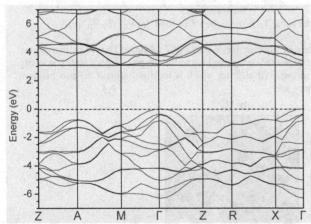

Figure 3. Band structure of rutile TiO_2 using the sX method. The red horizontal line marks the size of the band gap.

TiO$_2$ rutile	GGA	sX	Experiment
Band gap (eV)	1.86	3.1	3.1
Lattice constant (a/c) (Å)	4.64 / 2.97	4.56 / 2.98	2.59 / 2.96

Table 1. The band gaps and lattice constants calculated by the GGA and sX compared to the experimental value for rutile TiO$_2$.

Regarding the oxygen deficient, we consider firstly the oxygen vacancy of TiO$_2$. In this paper, we denote V_O^0, V_O^+ and V_O^{2+} for neutral, singly charged and doubly charged states respectively. The partial density of states (PDOS) calculated by sX were given in Figure 4. It reveals a defect level which is indicated by the arrow in the figure for V_O^0 and V_O^+. The position of the defect level is around 0.6 eV under the conduction band (CB) edge, agrees very well with experimental value [14]. There is no gap state for V_O^{2+}, because the excess electrons at the vacancy site would be neutralized by the doubly positive charge. The sX method gives a correct description of the localization of defect charge densities, which is not the case for the GGA [3-6].

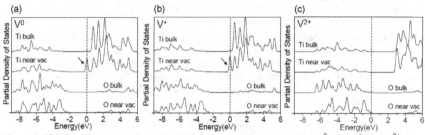

Figure 4. PDOS of O vacancy for rutile TiO$_2$ using the sX method, (a) V_O^0 (b) V_O^+ (c) V_O^{2+}.

Figure 5 shows a charge density contour for V_O^0 and V_O^+ of rutile TiO$_2$. The two excess electrons are located on two of three titanium atoms next to the oxygen vacancy site. The singly positive oxygen vacancy has an unpaired electron which is localized around the two titanium atoms close to the oxygen vacancy site.

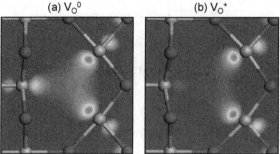

Figure 5. Charge density contour for the O vacancy site of rutile TiO$_2$, (a) V_O^0 (b) V_O^+, using sX method.

In the case of titanium interstitial, the DOS of Ti_i^0 is given in Figure 6.

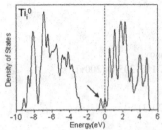

Figure 6. DOS of Ti interstitial neutral state (Ti_i^0) for rutile TiO_2 using the sX method.

Figure 7 shows the oxygen vacancy and titanium interstitial formation energy as a function of Fermi energy. In this figure, the red lines represent the oxygen vacancy and the black lines are the titanium interstitial which were both calculated by the sX method. It is found that the defect formation energies, for the neutral charge state, of oxygen vacancy and titanium interstitial are quite similar, 2.40 eV and 2.45 eV respectively, for an oxygen chemical potential in the O-poor condition. The similar size of these two calculated energies indicates that both are a cause of oxygen deficiency. The transition energy of the oxygen vacancy lies within the band gap, corresponds to the electrons trapped at adjacent titanium sites.

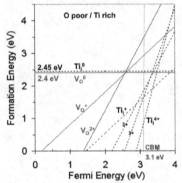

Figure 7. Formation energy against Fermi energy of the O vacancy and Ti interstitial in rutile TiO_2, calculated by sX.

CONCLUSIONS

This study presents an accurate band gap 3.1 eV for rutile TiO_2 by the sX exchange-correlation functional in DFT. The oxygen vacancy and titanium interstitial both create defect levels inside the band gap, and they are both a cause of oxygen deficiency according to their similar formation energy levels.

REFERENCES

1. S. J. Clark and J. Robertson, *Phys. Rev. B* **82**, 085208 (2010).
2. S. Wendt et al, *Science* **320**, 1755 (2008).
3. B. J. Morgan and G.W. Watson, *Surface Sci* **601**, 5034 (2007).
4. B. J. Morgan and G.W. Watson, *Phys. Rev. B* **80**, 233102 (2009).
5. C. DiValentin, G. Pacchioni and A. Selloni, *J. Phys. Chem. C* **113**, 20543 (2009).
6. S. Lany and A. Zunger, *Phys. Rev. B* **80**, 085202 (2009).
7. U. Diebold, *Surf. Sci. Rep.* **48**, 53 (2003).
8. S. J. Clark and J. Robertson, *Phys. Status Solidi B* **248**, 537 (2011).
9. L. J. Sham and M. Schluter, *Phys. Rev. Lett.* **51**, 1888 (1983).
10. J. P. Perdew and M. Levy, *Phys. Rev. Lett.* **51**, 1884 (1983).
11. R. W. Godby, M. Schluter and L. J. Sham, *Phys. Rev. Lett.* **56**, 2415 (1986).
12. D. M. Bylander and L. Kleinman, *Phys. Rev. B* **41**, 7868 (1990).
13. J. Robertson, K. Xiong and S. J. Clark, *Phys. Status Solidi B* **243**, 2054 (2006).
14. V. E. Henrich, G. Dresselhaus and H. J. Zeiger, *Phys. Rev. Lett.* **36**, 1335 (1976).

Mater. Res. Soc. Symp. Proc. Vol. 1352 © 2011 Materials Research Society
DOI: 10.1557/opl.2011.1079

Synergistic effects on band gap-narrowing in titania by doping from first-principles calculations: density functional theory studies

Run Long[1] and Niall J. English[1]
[1]The SEC Strategic Research Cluster and the Centre for Synthesis and Chemical Biology, Conway Institute of Biomolecular and Biomedical Research, School of Chemical and Bioprocess Engineering, University College Dublin, Belfield, Dublin 4, Ireland

ABSTRACT

The large intrinsic band gap in TiO_2 has hindered severely its potential application for visible-light irradiation. We have used a passivated approach to modify the band edges of anatase-TiO_2 by codoping of X (N, C) with transition metals (TM=W, Re, Os) to extend the absorption edge to longer visible-light wavelengths. It was found that all the codoped systems can narrow the band gap significantly; in particular, (N+W)-codoped systems could serve as remarkably better photocatalysts with both narrowing of the band gap and relatively smaller formation energies and larger binding energies than those of (C+TM) and (N+TM)-codoped systems. Our theoretical calculations help to rationalise experimental results and provide reasonably meaningful guides for experiment to develop more powerful visible-light photocatalysts.

INTRODUCTION

Titania (TiO_2)-based photocatalysts have received intense attention as promising photocatalytic materials [1]. However, their universal use is restricted to ultraviolent light ($\lambda <$ 385 nm) due to the wide band gap of titania (~3.2 eV for anatase). Further, photoexcited electron-hole pairs tend to recombine relatively easily in TiO_2. It is highly desirable to extend the optical absorption of TiO_2-based materials to the visible-light region with the minimum of photo-generated electron-hole recombination.

In general, doping is one of the most effective approaches to extend the absorption edge to the visible-light range. For instance, N-doped TiO_2 is considered to be a promising photocatalyst, and it has been investigated widely, both experimentally and theoretically [2]. However, due to strongly localized N 2p states at the top of valence band [3], the photocatalytic efficiency of N-doped TiO_2 decreases because isolated empty states trap an appreciable proportion of photo-excited electrons and reduce photo-generated current [4]. Besides N-doped TiO_2, C-doped TiO_2 also shows photocatalytic activity under visible-light [5].Conversely, transition metal doping can also promote photocatalytic efficiency, but this is hindered by the presence of carrier recombination centers and the formation of strongly localized d states in the band gap, which serves to reduce carrier mobility substantially [6]. Recently, Gai et al. [7] proposed using a passivated codoping approach, consisting of nonmetal and metal elements, to extend the TiO_2 absorption edge into the visible light range. Because defect bands are passivated, it is highly likely that they will be less effective as carrier recombination centers [8]. Recent experiments have reported that the addition of W to N-doped TiO_2 can increase photocatalytic activity under visible-light irradiation significantly [9, 10]. Our theoretical calculations on (N+W)-codoped anatase suggest that a continuum band is formed at the top of the valence band, and that W 5d orbitals locate below Ti 3d states at the bottom of conduction band, which narrows significantly the band gap and enhances visible light absorption [11].

THEORY

Spin-polarized DFT calculations were performed using the projector augmented wave (PAW) pseudopotentials as implemented in the Vienna ab initio Simulation Package (VASP) code [12, 13]. The Perdew and Wang parameterization [14] of the generalized gradient approximation (GGA) [15] was adopted for exchange-correlation. The electron wave function was expanded in plane waves up to a cutoff energy of 400 eV and a Monkhorst–Pack k-point mesh [16] of $4 \times 4 \times 4$ was used for geometry optimization [17, 18] and electronic property calculations. Both the atomic positions and cell parameters were optimized until residual forces were below 0.01 eV/Å. It is well known that GGA underestimates the band gap of TiO_2 significantly (2.0 eV vs. anatase 3.2 eV experimental value). Here, we include the on-site Coulomb correction for the Ti 3d states, the GGA + U method [19], which can improve the prediction of the band gap. Here, U = 5.3 eV for Ti, which is in agreement with the optimal value (5.5±0.5eV) [20, 21]; using this, the calculated band gap of pure anatase was 3.11 eV, agreeing well with the experimental value of 3.20 eV. A variety of U values (1.0, 2.0, 3.0, and 4.0 eV) were applied to the dopants. The calculated results for U=1.0 and 2.0 eV gave a qualitatively wrong metallic ground state as in the standard GGA calculations for W, and Os, and the once the U value was increased to 3.0 eV, the behavior of W- and Os-doped TiO_2 exhibited appropriate semiconductor characteristics. For Re-doped TiO_2, the band gap is not particularly dependent on the U value. Therefore, a moderate value of U = 3.0 eV has been applied to TM 5d states, similar to V-doped TiO_2 [22].

A relaxed $(2 \times 2 \times 1)$ 48-atom anatase supercell was used to construct doped systems. Single O and Ti atoms were replaced by single X (N, C) and TM (W, Re, Os) atoms, respectively. The codoped systems were created by simultaneous substitution of an O atom by an X (N, C) atom and a Ti atom by a TM (W, Re, Os) atom. In principle, the three 5d transition metal elements W, Re and Os served as n-type dopants and non-metal elements N, C as p-type dopants. It was found that formation of adjacent metal-nonmetal element pairs is more energetically favorable with respect to other configurations. For clarity, we have listed the total energies differences between X-TM nearest-neighbor (*1nn*), the second nearest-neighbor (*2nn*) and largest-distance (*Lnn*, one dopant in the center of the supercell with the other in a corner) doped systems in Table 1. Here we set the *1nn* total energies as zero eV. We will select the lowest-energy structures to examine their electronic properties.

Table 1. Total energy difference of different doping systems (in eV). The total energy of nearest-neighbor configuration (*1nn*) has been set as zero.

	N+W	N+Re	N+Os	C+W	N+Re	C+Os
1nn	0	0	0	0	0	0
2nn	0.48	0.95	1.0	1.1	0.35	3.1
Lnn	0.57	1.1	1.1	1.3	0.56	3.3

10

DISCUSSION

We propose to codope anatase with donor-acceptor pairs to reduce the formation energy and to narrow the band gap further. Therefore, we calculated the formation energy for (X+TM)-codoped system, *i.e.* (N+W), (N+Re), (N+Os), (C+W), (C+Re), and (C+Os), according to

$$E_{form} = E(TM@Ti + X@O) - E(TiO_2) - \mu_X - \mu_{TM} + \mu_O + \mu_{Ti} \qquad (1)$$

where $E(TM@Ti + X@O)$ is the total energy of the codoped system. The formation energies are also summarized in Table 1. The results show that codoping of X (N, C) with TM (W, Re, Os) reduces the formation energies significantly with respect to N and C monodoping under O-rich conditions, which corresponds to the usual growth conditions for synthesis samples in experiments. This indicates that codoping is beneficial for C or N introduction into the titania lattice. Hence, one could select W, Re, and Os to act as the codopants with N or C to favor the incorporation of N or C into the titania lattice in experiments. Furthermore, (N+TM)-codoped systems have lower formation energies than (C+TM)-doped TiO_2, which means that synthesis of the (N+TM) samples is relatively easier than the (C+TM) case.

Before we study the passivation effect on the band gap of TiO_2 via X-TM codoping, one needs to consider the stability of defect pairs. Therefore, we calculated the defect pair binding energy [33] according to

$$E_b = E(TM@Ti) + E(C@O) - E(TM@Ti + C@O) - E(TiO_2) \qquad (2)$$

$$E_b = E(TM@Ti) + E(N@O) - E(TM@Ti + N@O) - E(TiO_2) \qquad (3)$$

Positive E_b values indicate that the defect pairs tend to bind to each other and are stable. The calculated binding energies for the (C+W), (C+Re), (C+Os), (N+W), (N+Re), and (N+Os) pairs are 3.15, 0.65, 3.94, 2.64, 0.23, and 2.54 eV, respectively, indicating that C-W, C-Os, N-W, and N-Os impurities pairs are significantly stable relative to the isolated dopants. Furthermore, these binding energies are comparable with the reported result of Gai et al. [7]. However, the values of binding energy for C-Re and N-Re impurities pairs are not large enough and may possibly be broken up at high temperatures during the synthesis process. To confirm this, we also calculated the binding energies with a 108-atom supercell using a $2 \times 2 \times 2$ *k*-mesh. The corresponding values are 2.84, 0.50, 4.56, 2.28, 0.20, and 2.55 eV, respectively, indicating that the binding energies from the 48-atom system is reasonable. The large binding energy results arises from charge transfer from donor to acceptor and the strong associated Coulomb interaction is due to interactions between positively charged donors and negatively charged acceptors. The ELF is plotted in Fig. 1 at the (100) surface of bulk anatase for these three doped systems and the Bader charges are summarized in Tables 2 and 3 [34, 35]. Here, the partial optimized geometries are presented in Fig. 2 to compare with Tables 2 and 3. Fig. 1 shows that C-W and C-Re bonds exhibit ionic behavior while C-Os shows covalent characteristics. Table 2 shows that the C ion has a charge of -1.20 |e| for C@O monodoping, while it is -2.60 |e| for (C+W)-codoping, with more electrons transferring from the W and adjacent Ti atoms to the C ion. The bond length of C-W is 1.868 Å, which is shorter than that of the C-Ti length of 2.196 Å for C@O doping, indicating further that a strong C-W bond forms. For (C+Re), the optimized C-Re bond length is only 1.624 Å, shorter than that of C-W and significantly shorter than that of C-Ti. Therefore, a much stronger interaction between the C and Re ion takes place. For the (C+Os) system, the C-

Os bond behavior is different from C-W and C-Re (cf. Fig. 1) such that the formation of a covalent C-Os bond takes place through sharing of electrons between Os and C. The Bader charges also confirm this result, in which the C ion has a charge of -2.74 |e|. This method could be also employed to analyze the (N+TM) system.

Table 2: Average Bader Charges (|e|) on dopant atoms and their adjacent atoms in C-, and (C+W), (C+Re), and (C+Os)-doped TiO_2. Bond lengths of C-Ti and C-W, C-Re, and C-Os are also reported. The number in parenthesis denotes the number of nearest-neighbor atoms around a dopant.

	C-doped	(C+W)-doped	(C+Re)-doped	(C+Os)-doped
C	-1.20	-2.60	-2.60	-1.74
TM		4.82	2.94	0.97
O		-1.52(5)	-1.29(5)	-1.31(5)
Ti	2.69(3)	2.65 (2)	2.58(2)	2.71(2)
Bond length(Å)	2.196	1.868	1.624	1.743

Table 3: Average Bader Charges (|e|) on dopant atoms and their adjacent atoms in N-, and (N+W), (N+Re), and (N+Os)-doped TiO_2. Bond lengths of N-Ti and N-W, N-Re, and N-Os are also specified. The number in parenthesis denotes the number of nearest-neighbor atoms around a dopant.

	N-doped	(N+W)-doped	(N+Re)-doped	(N+Os)-doped
N	-1.41	-2.02	-1.94	-0.74
TM		4.22	2.68	2.55
O		-1.47(5)	-1.31(5)	-1.31(5)
Ti	2.74(3)	2.72 (2)	2.59(2)	2.75(2)
Bond length(Å)	2.026	1.851	1.717	1.798

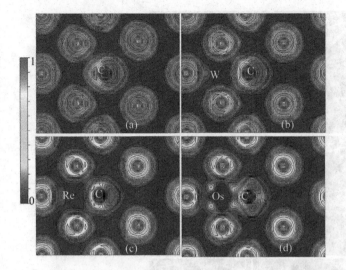

Figure 1: Electron localization function contour plots (ELF) on the (100) surface of bulk materials for (a) C-, (b) (C+W)-, (c) (C+Re), and (C+Os)-doped anatase.

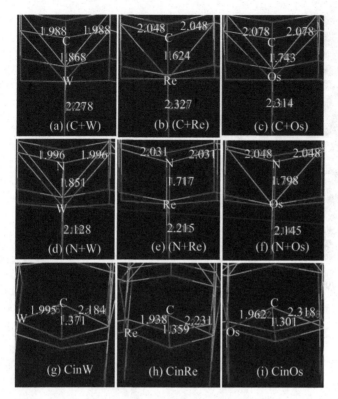

Figure 2: Geometry-optimized structures for (a) (C+W)-, (b) (C+Re), (c) (C+Os) (d) (N+W), (e) (N+Re), (f) (N+Os), (g) C in W, (h) C in Re, and (i) C in Os-doped anatase. The bond length unit is Å

Secondly, we have examined synergistic effects on narrowing the band gap of TiO_2 by codoping X (N, C) with TM (W, Re, Os). Hence, the calculated DOS and PDOS for these six systems, compared to the results with that of pure TiO_2, are shown in Fig. 3. It may be seen that the reduction in band gap caused by codoping is very significant with respect to that observed in the corresponding X (N, C) and TM (W, Re, Os) monodoped cases, leading to a large reduction of band gap in these doped systems.

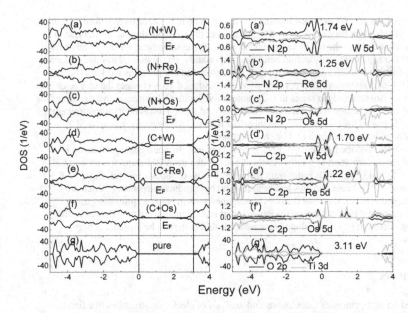

Energy (eV)

Figure 3: DOS of undoped anatase and (C+W), (C+Re), (C+Os), (N+W), (N+Re), and (N+Os)-codoped anatase and PDOS for impurity atoms (black for p orbital and green for d orbital) with 48-atom supercell. The top of the valence band of pure anatase is taken as the reference level. The dashed lines represent the Fermi level, E_F.

For the (N+W) codoped system (cf. Figs. 3 a&a'), identifying the formation of a continuous band of hybridized N2p-O2p states, which consists of the valence band edge and W5d orbitals dominating the conduction band edge, leading to a significant band gap reduction of about 1.37 eV [11]. It means (N+W) codoped TiO2 is a effective photocatalysts to absorb visible-light. Then, we examine the other two (N+TM)-codoped systems. (N+Re) (cf. Figs. 3b&b') and (N+Os) (cf. Figs. 3c&c') possess much smaller band gaps than those of both pure anatase and the (N+W) system. However, there are some shortcomings in them. Firstly, the binding energy of (N+Re) system is so small that it is easily departed during synthesis process. Secondly, some empty states located in the band gap of (N+Os) system that can act as recombination centres and reduce phonon absorption efficiency. Although these two codoped systems extend substantially the absorption edge to longer visible-light ranges, it's expected they are not effective and applied photocatalysts because of either unstable or low efficiency.

Compared to (N+TM)-codoped systems, (C+TM) exhibits much better effects on band gap narrowing, which may be due to deeper C acceptor energy levels than those of N, and a much stronger interaction between C and TM atoms. For the (C+W) codoped system (cf. Figs. 3d & d'), the band gap is reduced by about 1.41 eV with formation of isolated band above VBM with an value of 0.83 eV and CBM down-shifting by about 0.37 eV, which is different from the

(N+W) system. Like (N+W)-codoped titania [11], a C2p-W5d hybridized band is formed, which renders the hybridization between C 2p and O 2p stronger than C monodoping, giving rise to a fully occupied band above the top of valence band edge. At the same time, partial W 5d states are located at the edge of the conduction band, and importantly, the isolated states disappear and a continuum-like band is formed. This finding is very similar to our previous work on (N+Ta)-codoped TiO_2 [36]. Obviously, incorporation of W into C-doped TiO_2 changes the character of C 2p orbitals from isolated midgap states to C 2p states mixed with O 2p states above the top of the valence band and conduction band edge consisting of W 5d itself. However, its formation energy is nearly double that of the (N+W) case and therefore it would not be easily synthesized experimentally under either O-rich or Ti-rich conditions.

The (C+Re) codoped system (cf. Figs. 3 e&e') reduces the band gap significantly by about 1.89 eV, with formation of a fully occupied band above the VBM about 0.4 eV and CBM down-shifting by 1.43 eV, which is different from the (C+Os) codoped cases (cf. Figs. 3 f & f'). The band gap of (C+Os) case narrows less than (C+Re) but produces gap states, which are partially occupied. This implies that the (C+Os)-codoped system is not as promising as a possible photocatalytic material due to the existence of recombination centers. However, the small binding energy of (C+Re)-codoped anatase leads this system to be unstable at high temperatures.

CONCLUSIONS

Based on first-principles calculations and analysis of electronic structures, we find that although (C+TM) (TM=W, Re, Os)-codoped systems narrow the band gap to a slightly larger extent than those of (N+TM) systems, (N+TM) has more medium formation energies, which is of greater convenience for synthesis. Band gap narrowing is a highly important contribution to the enhancement of optical absorption under visible-light irradiation. The calculated results suggest that codoping of N or C with TM can effectively reduce the formation energy vis-à-vis N and C monodoping and enhance the N or C concentration. The results indicate that (N+W)-codoped anatase may serve as a very promising photocatalytic material because of its relatively small formation energy and large binding energy.

ACKNOWLEDGMENTS

This work was supported by the Irish Research Council for Science, Engineering and Technology (IRCSET). The authors thank Science Foundation Ireland and the Irish Centre for High End Computing for the provision of computational resources.

REFERENCES

1. M.R. Hoffmann, S.T. Martin, W.W. Choi and D.W. Bahnemann, Chem. Rev. 95, 69 (1995).
2. R. Asahi, T. Morikawa, T. Ohwaki, K. Aoki and Y. Taga, Science 293, 269 (2001).
3. H. Irie, Y. Watanabe and K. Hashimoto, J. Phys. Chem. B 107, 5483 (2003).
4. Z. Lin, A. Orlov, R.M. Lambert and M.C. Payne, J. Phys. Chem. B 109, 20948 (2005).
5. X.B. Chen and C.J. Burda, J. Am. Chem. Soc. 130, 5018 (2008).
6. W. Mu, J.M. Herrmann and P. Pichat, P. Catal. Lett. 3, 73 (1989).

7. Y.Q. Gai, J.B. Li, A.S. Li, J.B. Xia and S.H. Wei, Phys. Rev. Lett. 102, 036402 (2009).
8. K.S. Ahn, Y. Yan, S. Shet, T. Deutsch, J. Turner and M. Al-Jassim, Appl. Phys. Lett. 91, 231909 (2007).
9. B.F. Gao, Y. Ma, Y.A. Cao, W.S. Yang and J.N. Yao, J.Phys. Chem. B 2006, 110, 14391 (2006).
10. Y.F. Shen, T.Y. Xiong, T.F. Li and K. Yang, Appl. Catal., B 83, 177 (2008).
11. R. Long and N.J. English, Appl. Phys. Lett. 94, 132102 (2009).
12. G. Kresse and J. Hafner, Phys. Rev. B 47, 558 (1994).
13. G. Kresse and J. Furthermüller, J. Phys. Rev. B 54, 11169 (1996).
14. J.P. Perdew, K. Burke and M. Ernzerhof, Phys. Rev. Lett. 77, 3865-3868 (1996).
15. J.P. Perdew and Y. Wang, Phys. Rev. B, 45, 13244 (1992).
16. H.J. Monkhorst, H. J.; Pack, J. D. Phys. Rev. B 1976, 13, 5188-5192.
17. E.R. Davidson, Methods in Computational Molecular Physics edited by G.H.F. Diercksen
18. S. Wilson, Vol. 113 NATO Advanced Study Institute, Series C (Plenum, New York, 1983), p. 95.
19. S.L. Dudarev, G.A. Botton, S.Y. Savarsov, C.J. Humphreys and A.P. Sutton, Phys. Rev. B 57, 1505 (1998).
20. C.J. Calzado, A. Hernández and J.F. Sanz, Phys. Rev. B 77, 045118 (2008).
21. V.I. Anisimov, M.A. Korotin, I.A. Nekrasov, A.S. Mylnikov, A.V. Lukoyanov, J.L. Wang and Z. Zeng, J. Phys.: Condens. Matter 18, 1695 (2006).
22. X.S. Du, Q.X. Li, H.B. Su and J.L. Yang, Phys. Rev. B 74, 233201 (2006).
23. N.C. Wilson, I.E. Grey and S.P. Russo, J. Phys. Chem. C. 111, 10915 (2007).
24. C. Di Valentin, G. Pacchioni, A. Selloni, Chem. Mater. 2005, 17, 6656-6665.
25. E. Finazzi and C.Di Valentin, J. Phys. Chem. C 111, 9275 (2007).
26. K. Reuter and M. Scheffler, Phys. Rev. B 65, 035406 (2001).
27. CRC Handbook of Chemistry and Physics 87th ed.; Lide, D. R. Taylor & Francis: London, 2006.
28. C. Di Valelntin, G. Pacchioni, A. Selloni, S. Livraghi and E.J. Giamello, Phys. Chem. C 119, 11414 (2005).
29. B. Chi, L. Zhao, T. Jin, J. Phys. Chem. C 111, 6189 (2007).
30. M. Batzill, E.H. Morales and U. Diebold, Phys. Rev. Lett. 96, 026103 (2006).
31. K.S. Yang, Y. Dai, B.B. Huang and S. Han, J. Phys. Chem. B. 110, 24011 (2006).
32. A.D. Becke and K.E. Edgecombe, J. Chem. Phys. 92 5397 (1990).
33. J. Li, S.H. Wei, S.S Li and J.B. Xia, Phys. Rev. B 74, 081201 (2006).
34. G. Henkelman, A. Arnaldsson and H. Jónsson, Comput. Mater. Sci. 36, 354 (2006).
35. E. Sanville, S.D. Kenny, R. Smith and G. Henkelman, J. Comp. Chem. 28, 899 (2007).
36. R. Long and N.J. English, Chem. Phys. Lett. 48, 175 (2009).

Synthesis

Mater. Res. Soc. Symp. Proc. Vol. 1352 © 2011 Materials Research Society
DOI: 10.1557/opl.2011.759

On the Sol-gel Synthesis and Characterization of Titanium Oxide Nanoparticles

Varun Chaudhary, Amit K. Srivastava and Jitendra Kumar

Materials Science Programme, Indian Institute of Technology Kanpur, Kanpur, India
Email- jk@iitk.ac.in

ABSTRACT

TiO$_2$ nanoparticles have been prepared by sol-gel process using titanium isopropoxide as a precursor with ethanol and water as solvents. The synthesis involves gel formation, digestion for 24h, drying at 100°C for 10h, and calcination in air at 500-800°C for 2h. The resulting powder has been studied with respect to phase(s), morphology, optical absorption and photo - luminescence (PL) behaviour. The calcination of dried sol-gel product at 500°C for 2h leads to formation of anatase phase that possesses a tetragonal structure (a = 3.785 Å, c = 9.514 Å, Z = 4), average crystallite size ~ 11 nm and band gap of 3.34 eV. Further, increasing the time (t) of calcination causes crystallite growth that follows the relation d = α – β exp (-t/τ), α = 18.1 nm, β = 9.6 nm and τ = 6.9h. However, calcination of sol-gel product at 800°C for 2h gives rise to a rutile phase (tetragonal a = 4.593Å, c = 2.959Å, Z = 2), average crystallite size ~ 25 nm and band gap of 3.02 eV. The anatase phase exhibits strong PL emission peaks (excitation wavelength 405 nm) at 2.06 and 1.99 eV due to defect levels within the energy band gap. This observation has been attributed to finite size effects occurring in nanoparticles.

INTRODUCTION

Titanium dioxide (TiO$_2$) exhibits unusual structural, optical, electronic, magnetic and chemical properties [1, 2]. As a consequence, it has found wide applications in pigments, UV protection creams, photo-catalysis, solar cells, water and air purification, synthesis of inorganic membranes, etc. [3, 4]. Its important phases include anatase, rutile (both tetragonal, described by sharing of (TiO$_6$)$^{2-}$ octahedra), brookite (orthorombic), and a recently discovered oxygen deficient λ-Ti$_3$O$_5$, having a monoclinic crystal structure and photo-reversible characteristics [5, 6]. In bulk, anatase and brooklite are less stable and undergo transition to rutile phase which remains thermodynamically stable under atmospheric pressure up to the melting point (1840°C). The transformation to rutile phase is dependent on several parameters such as particle size, temperature and the environment. TiO$_2$ thin films have found application in dye-sensitized solar cells (DSSC) because of their interconnected pore networks and a large surface area, which allows sufficient dye adsorption and efficient light harvesting. Hence, the performance of such cells depends on the nature of porous structure and average particle size besides the phase. The conversion efficiency of ~11% or more is already attained in the laboratory [7]. However, the commercial devices have yet to reach at that level. Cell stability is another issue requiring attention. One way to improve the efficiency and stability of DSSCs may be through the exploitation of electrodes with TiO$_2$ nanoparticles of superior characteristics [8]. The short-circuit photocurrent of the rutile phase-based DSSC is reported to be about 30 % lower than that of cell produced with TiO$_2$ anatase [9]. The electron transport is also slower in the rutile layer due to nature of its interparticle connectivities. Nevertheless, high dielectric constant and refractive index make rutile phase useful in other applications [10]. Obviously, the potential of titania is strongly dependent on the crystal structure, morphology and average size of the particles. It is therefore necessary to develop a method which can precisely control the size and shape of the titania particles of a particular phase. In this work, an attempt has been made to i) synthesize highly stable titania nanoparticles of controllable size and ii) characterize them with regard to phase(s), morphology, optical absorption and photoluminescence. Various techniques like aerosol, wet chemical, micro emulsion and sol-gel synthesis are used for the production of titanium oxide powders [11]. The sol-gel process offers several advantages, viz., low processing temperature, better homogeneity, high purity, accurate control of composition and cost [12]. Hence, sol-gel process has been used here taking diethenol amine as an additive.

EXPERIMENTAL

The titanium dioxide powder was prepared by sol-gel process using titanium isopropoxide, $[Ti(OC_3H_7)_4]$, as a precursor and ethanol as a solvent. After dissolution of 11.2 mL of $Ti(OC_3H_7)_4$ in ethanol, a solution of 0.73 ml water in ethanol was mixed and 1 mL diethenol amine $[C_4H_{11}NO_2]$ added dropwise under continuous stirring for 2h to realize a transparent sol. $Ti(OC_3H_7)_4$: C_2H_5OH: H_2O: $C_4H_{11}NO_2$ molar ratio was set at 4:140:4:1. The sol was digested for 24h and dried subsequently at $100^{\circ}C$ for 10h, calcined for 2h elevated temperatures, and cooled at a rate of ~ $8.3^{\circ}C/min$. Thermogravimetric analysis (TGA) of the dried sol-gel product was carried out by raising its temperature at a rate of $4^{\circ}C/min$ from 50 to $850^{\circ}C$ in air to ascertain the conditions of TiO_2 formation. While an X-ray diffractometer (Thermo Electron ARL X' TRA) with the CuKα radiation was used for identification of phase(s) and determining the average crystallite size, fourier transform infrared (FTIR) spectrometer (BRUKER Vertex-70) was employed for ascertaining the bonds and stretching modes present. A field emission scanning electron microscope (Carl Zeiss modal Supra 40VP) and UV-Vis- IR spectrophotometer (Varian, model Cary 5000) were used for observing the morphology and optical absorption behaviour, respectively. A spectrofluorophotometer (Edinburg Instruments model FLSP 920) was utilized to study the photoluminescence characteristics of the final product.

RESULT AND DISCUSSION

Figure 1 shows the weight (W) % versus temperature (T) and – (dW/dT) versus T plots of the dried sol-gel product when the temperature is raised at a rate of $4^{\circ}C$ per minute from 50 to $850^{\circ}C$. Accordingly, the total weight loss occurs in several steps and finally becomes ~55.8% at $553^{\circ}C$. Thus, the end product weighs ~ 44.2% of the original dried gel. The formation of gel involves hydrolysis and condensation. In hydrolysis, alkoxide group is replaced and hydroxyl metal alkoxide results which, in turn, produces polymerizable species and water / alcohol during the condensation process. The nature of the product however depends upon the additives (e.g., acetic acid, HNO_3, amines or HCl), amount of water, and the rate of mixing. Considering the TG data (i.e., 44.2% yield and nature of weight loss in different temperature intervals), polymerizable species possibly of $\{OCH (CH_3)_2\}_2Ti-O$ emerged during the condensation process. In fact, the polymerizable species contain just ~ 9.9 wt% of water (which gets liberated below $160^{\circ}C$) and ~ 44 wt% of titanium oxide, formed above $550^{\circ}C$ (Figure 1). Decomposition involves evolution of water and carbon or its compounds (e.g., CO, CO_2) consuming atmospheric oxygen in steps. The balanced decomposition reaction can be written as

$$\{OCH(CH_3)_2\}_2Ti\text{--}O + (6+y/2)O_2 \rightarrow TiO_2 + 7H_2O + (6-y)CO + yCO_2$$

where y lies between zero and six.

Figure 2 presents the FTIR spectra of dried gel before and after decomposition at 500°C for 2h. The broad peaks at 3367 and 2558 cm^{-1} observed in dried gel sample (figure 2a) have signatures of hydroxyl and carboxylic group, respectively. Also, the sharp peaks centred on 1621, 1451, and 1080 cm^{-1} can be attributed to C=C (alkenes) stretching, -C-H (methyl or methylene) bending and –C-O stretching, respectively [13]. A broad peak at 736 cm^{-1} arises due to =C-H bending [13]. After decomposition of gel at 500°C for 2h, the product depicts a major peak at 506 cm^{-1} (figure 2b) corresponding to Ti-O bond [14]. Another peak at ~3367 cm^{-1} arises due to the adsorbed water after the decomposition.

XRD patterns of powders obtained by decomposition of dried gel samples for 2h each at 500, 600, 700 and 800°C are presented in figure 3. The resulting product at 500°C corresponds to anatase phase of TiO_2 having tetragonal structure with lattice parameter a = 3.785 Å, c = 9.514 Å, Z = 4 and space group $I4_1/amd$ (JCPDS # 21-1272). As the calcination temperature is raised to 600°C, the diffraction peaks of anatase phase improve slightly and become sharp as well. In addition, the some extra peaks appear possibly due to emergence of rutile phase (as discussed later). With increase of the calcination temperature to 800°C, a single phase emerges (Figure 3) which matches well with rutile TiO_2 having a tetragonal structure and parameters a = 4.593Å, c= 2.959Å, Z = 2, and space group $P4_2/mnm$ (JCPDS # 21-1276). The weight fractions of the two phases (f_A and $f_R = 1- f_A$) for TiO_2 samples produced via calcination of dried sol-gel product at 600 and 700°C have been determined using intensities of 101 and 110 diffraction peaks of anatase (I_A) and rutile (I_R), respectively and the formula $f_A =1/(1+1.26\times I_R/I_A)$ [17]. The weight percent of rutile phase is found to increase from 45.2 to 72 when the calcination temperature raised from 600 to 700°C (table1). It can be noticed that the diffraction peaks of anatase phase are suppressed while that of rutile improved progressively with rise in calcination temperature (Figure 3). The average crystallite size (d) of phase(s) has been deduced from the Scherrer formula, taking the diffraction peaks mentioned above in each case, $d = 0.89 \lambda / B_s \cos \theta$,

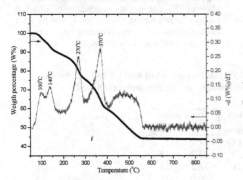

Figure 1 Weight percentage (W%) versus temperature (T) and –[d(W%)/dT] versus T plots of dried sol–gel product.

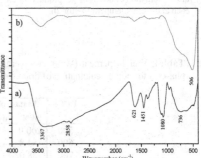

Figure 2 FTIR spectra of dried a) sol-gel and b) TiO_2 obtained by decomposition of gel at 500°C for 2h.

23

where λ is the X-ray wavelength, θ is the Bragg angle and B_s is the corrected full width at half maximum of peak taking silicon as standard [15]. The weight percent and average crystallite size of each phase are summarized in table 1. In order to understand the growth of anatase and rutile phases, dried sol-gel product was calcined separately at 500 and 800°C for 2 to 24h and average crystallite size determined as before. Interesting, the average crystallite size for anatase phase increased from 11 to 18nm when the calcination time increased from 2 to 24h at 500°C (inset of figure 4). The growth follows the relation $d = \alpha - \beta \exp(-t/\tau)$, where constants α, β and τ are 18.1nm, 9.6nm and 6.9h, respectively. 'τ' defines the time in which particles attains 80% of its ultimate size. However, for rutile, average crystallite increases up to 4h of calcination time at 800°C and no growth occurs thereafter. Figure 4 shows the XRD patterns of anatase and rutile phases resulting after calcination of dried gel for 24h each at 500 and 800°C, respectively.

Figure 5 shows the scanning electron micrographs of the TiO_2 powder obtained by calcination of the dried sol-gel product for 2h each at 500, 700 and 800°C which belonging to anatase, mixed and rutile phases, respectively. While the anatase phase (Figure 5a) consists of

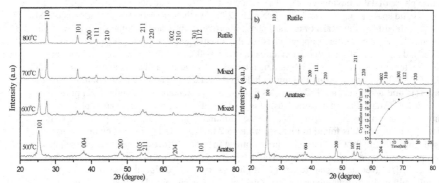

Figure 3 XRD patterns of TiO_2 powder formed after calcination of dried sol-gel product for 2h each at different temperature for 2h.

Figure 4 XRD patterns of a) anatase and b) rutile phase of TiO_2 emerged after calcination of sol-gel product for 24h each at 500 and 800°C with the inset of the variation of average crystallite size (d) as a function of dried sol-gel calcination time at 500°C.

Table1. Weight percent (W%) and average crystallite size (d) of anatase and / or rutile phase(s) formed by calcination of dried sol-gel product at different temperatures (T).

T (°C)	Anatase phase		Rutile phase	
	W%	d (nm)	W%	d (nm)
500	100.0	11	-	-
600	54.8	20	45.2	22
700	28.0	21	72.2	23
800	-	-	100.0	25

different shape particles of various sizes, rutile phase (Figure 5c) comprises of irregular platelets stacked like steps. The mixed phase (Figure 5b) depicts nearly uniform distribution of individual crystallites with agglomerates as well.

Figure 6 shows the optical absorption spectra of anatase and rutile phases of titanium dioxide powders in the wavelength range of 200 – 750 nm. The anatase phase exhibits moderate absorption above 400 nm, optical transition in the 335 – 395 nm intervals and high absorption below 335 nm. The optical transition zone gets red shifted whereas absorption higher above 425 nm and lower below 350 nm in rutile vis - a - vis anatase sample. The energy band gap has been determined by inflection point in dA/dλ verses λ plot (inset of Figure 6). The values found are 3.34 and 3.02 eV for anatase and rutile phase, respectively. The corresponding values for bulk TiO_2 are 3.18 and 3.03 eV [16]. The blue shift in the band gap for anatase phase is caused by quantum size effects as observed by Kumar et al. in TiO_2 films [17].

Figure 7 shows the photoluminescence spectra of TiO_2 phases recorded with the excitation wavelength of 405 nm. The featurs of PL spectra can be summarized as: i) a broad and weak emission peak at 450 nm (or 2.75 eV), two strong and sharp emissions at 601 nm (or 1.06 eV).

Figure 5 SEM images of dried sol gel after calcination for 2h at a) 500, b) 700 and c) 800 °C.

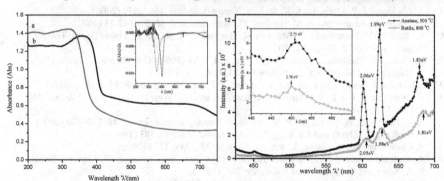

Figure 6 Optical absorption spectra of TiO_2 powders obtained by calcination of the dried sol-gel product for 2h at a) 500°C and b) 800°C. The inset shows the corresponding differential absorption curves.

Figure 7 Photoluminescence (PL) spectra of TiO_2 powders obtained by calcination of the dried sol-gel product for 2h at a) 500°C and b) 800°C. The inset shows zoom portion of wavelength range 440-465 nm.

and 624 nm (or 1.99 eV), and another moderate peak at 650 nm (or 1.83 eV) for anatase TiO_2. ii) all the above emission peaks continue to be present in rutile phase but are somewhat broad, very weak, and red shifted (i.e., towards lower energy). These results are at variance with the observation of Jin et al.[18], who observed no PL emission in anatase TiO_2 films but strong peaks (energy range 1.91 – 2.28 eV) with addition of $ZnFe_2O_4$ and attributed them to impurity - or oxygen vacancies - related centres. The strong emission peaks seen in anatase phase here are arising due to defect energy levels created within the band gap because of oxygen vacancies and /or finite size effects (i.e., small crystallite size ~ 11 nm and associated stress). Moreover, the origin of PL emission in the two phases lies with the defect centres, being less in case of rutile due to higher crystallite size (~ 25 nm).

CONCLUSIONS

TiO_2 nanoparticles of anatase phase with controllable average size (d) 11-18 nm can be synthesized by adjusting the time of calcination at 500°C of the dried sol – gel product. The particle growth follows the relation $d = \alpha - \beta \exp(-t/\tau)$. Its PL emission with excitation wavelength of 405 nm arising at 1.83, 1.99 and 2.06 eV can be attributed to the defect energy levels created within the band gap due to oxygen vacancies and / or finite size effects. Pure rutile phase formation requires calcination temperature of 800°C and yield particles of 'd' = 25 nm.

REFERENCES

1. M. R. Hoffmann, S. T. Martin, W. Choi, and D. W. Bahnemann, *Chem. Rev.* **95**, 69–96 (1995).
2. A. C. Pierre, *Ceramic Bulletin* **70**, 1281–1288 (1991).
3. U. Diebold, *Surface Science Reports* **48**, 53-229 (2003)
4. A. I. Kingon, J. P. Maris and S. K. Steiffer, *Nature* (London) **406**, 1032 (2000)
5. J. Muscat, V. Swamy and N. M. Harrism, *Physical Review B* **65**, 224112-15 (2002)
6. S. Ohkoshi, Y. Tsunobuchi, T. Matsuda, K. Hashimoto, A. Namai, F. Hakoe and H. Tokoro, *Nature Chemistry* **2**, 539-545 (2010).
7. M. Grätzel, J. Photochem. Photobio. A Chem, **164**, 3–14 (2004); C R Chimie **9**, 578–83 (2006)
8. A. Hagfeldt and M. Grätzel, *Acc. Chem Res* **33**, 269–77 (2000).
9. N.G. Park, J. van de Lagemaat and A. J. Frank, *J. Phys. Chem. B* **104**, 8989- 8994 (2000).
10. W. Wang, B. H. Gu, L. Y. Liang, W. A. Hamilton and D.J. Wesolowski, *J. Phys. Chem. B* **108**, 14789–92 (2004)
11. A. K. Bhanwala, A. Kumar, D. P. Mishra and J. Kumar, *Aerosol Science* **40**, 720–730 (2009).
12. C. C. Wang, Z. Zhang and J.Y.Ying, *Nano Structured Mater.* **9**, 583 (1997)
13. A. Chowdhury and J. Kumar, *Mater. Sci. Technol.* **22**, 1249–1254 (2006).
14. A. Amlouk, L. Mir, S. Kraiem and S. Alaya, *J. Phys. and Chem. of Solids* **67**, 1464–68 (2006)
15. R. A. Spurr and H. Myers. *Anal. Chem.* **29**, 760, (1957)
16. S. Kitazawa, Y. Choi and S. Yamamoto, *Vacuum* **74**, 637 (2004)
17. P. Madhu Kumar, S. Badrinarayanan, Murali Sastry, *Thin solid films* **358**, 122-130, (2000)
18. Y. Jin, G. Li, Y. Zhang and L. Zhang, *J. Phys. D: Appl. Phys.* **35**, 37-40 (2002)

Mater. Res. Soc. Symp. Proc. Vol. 1352 © 2011 Materials Research Society
DOI: 10.1557/opl.2011.1053

Slow Aggregation of Titania Nanocrystals in Acidic Hydrosols

Olga Pavlova-Verevkina[1], Ludmila Ozerina[1], Natalia Golubko[1] and Abdelkrim Chemseddine[2]

[1]Karpov Institute of Physical Chemistry, Obuha side street 3-1/12, building 6, 105064
Moscow, Russia
[2]Helmholtz-Zentrum Berlin fur Materialien und Energie GmbH, Hahn-Meitner-Platz 1,
D-14109 Berlin, Germany

ABSTRACT

The kinetics of slow aggregation of monodisperse TiO_2 nanocrystals in the acidic
hydrosols at room temperature was studied for months by turbidimetry. The dependence of
the initial rate of aggregation on the pH was calculated. The comparison of results obtained by
turbidimetry and small angle X-ray scattering permits to suppose that very loose aggregates
form at the low pH in HCl solution. The dependencies obtained in this work for room
temperature can be taken into consideration at the tuning of TiO_2 nanoparticles morphology
through thermal treatment of hydrosols.

INTRODUCTION

Aqueous dispersions of TiO_2 nanoparticles (hydrosols) stabilized under acid conditions
are generally used as precursors for the processing of nanostructured materials [1-4]. Protons
of acid adsorb on TiO_2 nanoparticles and the originating double electrical layer stabilizes
particles and inhibits aggregation. On the other hand, an increase of concentration of acid or
other electrolytes leads to compression of the double layer that destabilizes sols [4-5]. At
critical concentrations of electrolytes C_c the sols fast coagulate, while at relatively low
concentrations of electrolytes aggregation is very slow. The morphology of aggregates
forming in the course of slow aggregation depends on structure of primary TiO_2 nanocrystals,
electrolytes concentration, pH, time of aggregation, and other parameters. Investigation of the
kinetics of slow aggregation could help in the tuning of TiO_2 nanoparticles morphology.

Our research efforts are focused on the kinetics of slow aggregation of monodisperse TiO_2
hydrosols stabilized by strong acids [6-8]. Evolution of turbidity spectra of the diluted sols
destabilized by additions of HCl and KCl was investigated for months. (KCl was used
because this electrolyte does not adsorb on TiO_2 surface.) This investigation revealed that the
slow structural changes in the sols were mainly due to aggregation of primary nanocrystals
but not due to Ostwald ripening. The initial nonaggregated sols had the Rayleigh light
scattering. In the course of aggregation the optical density gradually increased and the
character of turbidity spectra quantitatively changed.

In this paper, new results on influence of medium composition on kinetics of the slow
aggregation are presented.

EXPERIMENTAL DETAILS

Nearly monodisperse TiO_2 nanocrystals were obtained by stepped coagulation of polydisperse TiO_2 hydrosols through additions of hydrochloric acid [6]. The isolated narrow fractions of nanocrystals were mixed with certain quantities of water that resulted in the formation of stable sols with pH=1.

The stable sols were destabilized either by additions of HCl and KCl with concentration below C_c or by increase of the pH with the help of dilution by water. This induced slow aggregation that was studied for months using turbidimetry and small angle X-ray scattering (SAXS). All experiments were carried out at room temperature. The samples were kept in hermetic vessels and carefully agitated before each measurement.

Thermogravimetric analysis was used to determine TiO_2 concentration in the sols. High resolution transmission electron microscopy (HRTEM) [9] and powder X-ray diffraction (XRD) [8] were used to characterize nanocrystals and their aggregates. Turbidity spectra $D(\lambda)$ (D – optical density, λ - wavelength) were registered by a Shimadzu-UV 2501 PC spectrophotometer using cuvettes with thickness from 1 to 50 mm. Dependencies $D(t)$ were constructed from the obtained turbidity spectra for λ=const (t - time of aggregation). The values C_c were determined from the dependencies $D(C)$ obtained at t=0 (C - concentration of the added electrolyte) [6]. SAXS curves were registered using a KRM-1 camera with a slit collimation system, giving an overall range of momentum transfer of $0.07 < q < 4.26$ nm^{-1} [7]. A mean radius of gyration of particles R_g was calculated from the SAXS curves by the Guinier approximation [10].

RESULTS AND DISCUSSION

HRTEM and XRD revealed that the isolated TiO_2 fractions contained mainly nanocrystals of anatase modification. The nanocrystals were platelets of irregular shape that can be seen from Figure 1. Thicknesses of the platelets were rather similar and averaged 2-3 nm. Lateral sizes of the platelets ranged in the fractions from 4 to 40 nm, increasing with the decrease of C_c.

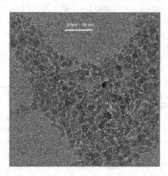

Figure 1. TEM of titania nanocrystals.

To study the slow aggregation process fractions with the smallest anatase nanocrystals were used. The stable sols prepared on basis of these fractions had R_g=6-9 nm.

The long-term turbidimetry investigation revealed that sols of several compositions aggregated at two stages. This is clearly seen from Figure 2 which shows kinetic dependencies D(t) obtained for sol of the following composition: m=0.5 mass.% TiO_2, C=0.5 M KCl, pH=2. Curves 1-3 correspond respectively to λ 580, 700 and 850 nm. These dependencies reveal that at t=1.7 month a noticeable acceleration of aggregation occurred and the second stage of aggregation began.

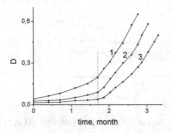

Figure 2. The kinetics of variation in the optical density of sol of the following composition: m=0.5 mass.% TiO_2, C=0.5 M KCl, pH=2. λ: 580 (1), 700 (2) and 850 nm (3). The dotted line indicates the beginning of the second stage of slow aggregation.

At the initial period of slow aggregation the kinetic dependencies D(t) are linear. It was shown that, at this stage, the dependencies D(t)/m depend on TiO_2 concentration m very slightly. This can be seen from Figure 3 which presents the dependencies D(t)/m of two sols, in which TiO_2 concentration differed in two times and the medium composition was identical.

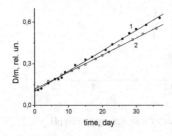

Figure 3. The kinetics of variations in the optical density of two sols with m 0.5 (1) and 1 (2) mass.% TiO_2, C=0.5 M/l KCl and pH=2. λ=520 nm.

So the influence of medium composition on kinetics of aggregation can be studied at different TiO_2 concentration.

The influence of the pH on the kinetics of slow aggregation was studied for a series of sols with different HCl and TiO_2 content. The pH of the sols was varied in the range 0.3-3. TiO_2 concentration was 0.1-3 mass. %. Figure 4 presents kinetics dependencies D(t)/Lm for four sols (L - thickness of cuvettes).

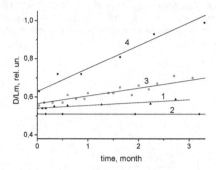

Figure 4. The kinetics of variations in the optical density of sols with pH 0.7 (1), 1.1 (2), 1.6 (3) and 3.0 (4). λ=420 nm.

From the linear dependencies obtained for the given series of sols values of initial rate of slow aggregation $V_0=kL^{-1}m^{-1} \cdot dD/dt$ were calculated (k - coefficient) and the dependence V_0(pH) was constructed (Figure 5). One can see from Figure 5 that V_0 monotonically increases both at the decrease of pH from 0.9 to 0.3 and at the increase of pH from pH 1.3 to 3. The sols with pH=0.9-1.3 are the most stable.

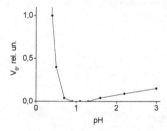

Figure 5. The dependence of initial rate of slow aggregation on pH.

The acceleration of aggregation at the decrease of pH from 0.9 to 0.3 may be due to compression of the double electric layer around nanocrystals with the increase of ionic strength [5]. On the other hand, the acceleration of aggregation at the increase of pH from 1.3 to 3 is probably connected with the reducing of surface charge on nanocrystals.

The slow aggregation of nanocrystals in HCl solution with pH=0.5 was studied using both turbidimetry and SAXS. Figure 6 shows turbidity spectra (a) and SAXS curves (b) registered in 1 (1) and 7 months (2) after acidulation of the initial sol.

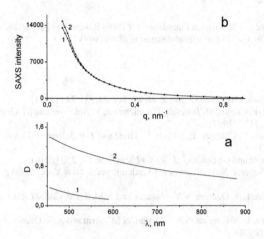

Figure 6. Turbidity spectra (a) and SAXS curves (b) of the sol with pH=0.5 and m=1 mass.% TiO$_2$ registered in 1 (1) and 7 months (2) after destabilization of the initial sol.

It can be seen from this figure that while the turbidity increased intensively with time, the SAXS curves changed very little. Correspondingly R$_g$ practically did not change. This permit to suppose that in the course of aggregation the light scattering occurred mainly from aggregates, while the SAXS occurred mainly from primary nanocrystals. That is the primary nanocrystals inside aggregates are separated from each other by rather thick interlayers of the medium.

So the aggregates forming slowly at pH=0.5 seem to be very loose particles. Obviously at such low pH values the surface charge on primary nanocrystals is very high that prevents close approaching of particles.

On the other hand, it was found earlier [7-8] that in the course of slow aggregation of two other sols, which had pH=2 and were destabilized by additions of KCl, R$_g$ increased noticeably. This permit to assume that the density of TiO$_2$ aggregates forming in the acidic hydrosols depends on the medium composition.

CONCLUSIONS

The kinetics of slow aggregation of monodisperse TiO$_2$ nanocrystals in the acidic hydrosols at room temperature was studied. The dependence of the initial rate of aggregation on the pH was calculated. The comparison of the results obtained by turbidimetry and SAXS permits to suppose that very loose aggregates form at the low pH in HCl solution. The dependencies obtained in this work for room temperature can be taken into consideration at the tuning of TiO_2 nanoparticles morphology through thermal treatment of hydrosols.

ACKNOWLEDGMENTS

The work was supported by the Russian Foundation for Basic Research (11-03-00292). The authors thank Prof. A.N. Ozerin for helpful discussion of this work.

REFERENCES

1. M.A. Anderson, M.J. Gieselmann and Q. Xu, *J. Memb. Sci.* **39**, (3), 243 (1988).
2. C.J. Barbe, F. Arendse, P. Comte, M. Jirousek, F. Lenzmann, V. Shklover and M. Gratzel, *J. Am. Ceram. Soc.* **80**, 3157 (1997).
3. A. Pottier, S. Cassaignon, C. Chaneac, F. Villain, E. Tronc and J.-P. Jolivet, *J. Mater. Chem. A.* **13**, 877 (2003).
4. P. Alphonse, A. Varghese and C. Tendero, *J. Sol-Gel Sci. Tech.* **56**, 250 (2010).
5. C.J Brinker and G.W. Scherer, *Sol-Gel Science,* (Academic press, New York, 1990) p. 239.
6. O.B. Pavlova-Verevkina, L.A. Ozerina, V.V. Nazarov and N.M. Surin, *Colloid Journal* **69**, 492 (2007).
7. O.B. Pavlova-Verevkina, L.A. Ozerina, S.N. Chvalun, N.M. Surin and A.N. Ozerin, *J. Sol-Gel Sci. Tech.* **45**, 219 (2008).
8. O.B. Pavlova-Verevkina, L.A. Ozerina, E.D. Politova, N.M. Surin and A.N. Ozerin, *Colloid Journal* **71**, 529 (2009).
9. A. Chemseddine and T. Moritz, *Eur. J. Inorg. Chem.* 235 (1999).
10. L.A. Feigin and D.I. Svergun, *Structure Analysis by Small-Angle X-ray Scattering and Neutron Scattering,* (Plenum Press, New York, 1987).

Mater. Res. Soc. Symp. Proc. Vol. 1352 © 2011 Materials Research Society
DOI: 10.1557/opl.2011.1008

Fabrication of Three-Dimensionally Ordered Macro-/Mesoporous Titania Monoliths by a Dual-Templating Approach

Zhiyan Hu,[1] Zhongjiong Hua,[1] Shaohua Cai,[1] Jianfeng Chen,[1] Yushan Yan,[2] and Lianbin Xu[1,2,*]
[1] Key Lab for Nanomaterials, Ministry of Education, Beijing University of Chemical Technology, Beijing 100029, China
[2] Department of Chemical and Environmental Engineering, University of California at Riverside, Riverside, CA 92521
E-mail: lbxu99@gmail.com

ABSTRACT

Three-dimensionally ordered macro-/mesoporous (3DOM/m) TiO_2 monoliths were fabricated by a dual-templating synthesis approach employing a combination of both colloidal crystal templating (hard-templating) and surfactant templating (soft-templating) techniques. Titania precursor, consisting of amphiphilic triblock copolymer Pluronic P123 as a mesopore-structure-directing agent and titanium tetraisopropoxide as a titanium source, was infiltrated into the void spaces of the poly(methyl methacrylate) (PMMA) colloidal crystal monolith. Subsequent thermal treatment produced 3DOM/m TiO_2 monolith. The macropore walls of the prepared 3DOM/m TiO_2 exhibit a well-defined mesoporous structure with narrow pore size distribution, and the mesopore walls are composed of nanocrystalline anatase TiO_2. The material also shows a high surface area (171 m^2/g), and large pore volume (0.402 cm^3/g).

INTRODUCTION

Due to its unique physicochemical properties, titania (TiO_2) has been extensively investigated for applications in photocatalysis, solar cells, electrochromic devices, and sensors [1]. The properties of TiO_2 strongly depend on its phase composition, crystallinity, and microstructure [1, 2]. Ordered mesoporous TiO_2 materials with a crystalline framework, high specific surface area and tailored pore structure are of much interest owing to their high reactivity, enhanced adsorption and sensing ability, and molecular sieving effect. The surfactant templating method through cooperative self-assembly of titania and surfactant has been well established for the fabrication of ordered mesoporous TiO_2 materials [3-5]. Recently, TiO_2 materials with hierarchically macro-/mesoporous structure have received considerable attention since they combine the advantages of efficient mass transport from macropores and high surface area from mesopores [6, 7]. Various synthesis strategies have been developed to fabricate hierarchically macro-/mesoporous TiO_2 [6]. The dual-templating synthesis approach employing a combination of both surfactant and colloidal crystal templating techniques offers a promising way for the fabrication of TiO_2 with three-dimensionally (3D) ordered macropores and mesopores [8, 9]. An additional advantage of 3D ordered macro-/mesoporous (3DOM/m) TiO_2 materials is that they exhibit enhanced light harvesting efficiency for photocatalytic and photovoltaic applications due to the multiple scattering and slow photon effects relating to their unique inverse opal photonic crystal structure [10, 11]. However, there have been few studies conducted on the production of 3DOM/m TiO_2 materials due to the high reactivity of hydrolysis and condensation of titania precursors. Very recently, Zhao et al. synthesized 3DOM/m TiO_2 films by a sol-gel dip-coating method [9]. 3DOM/m TiO_2 monoliths are also technologically important because of their ease of handling and potential applications in catalysis and separation [12, 13]. In this work, we report for the first time the fabrication of 3DOM/m TiO_2 monoliths by using poly(methyl methacrylate) (PMMA) colloidal crystals as hard templates and amphiphilic triblock copolymer Pluronic P123 ($EO_{20}PO_{70}EO_{20}$) as a mesopore-structure-directing agent (soft template).

EXPERIMENTAL

Preparation of PMMA colloidal crystal monoliths. Monodispersed PMMA spheres were synthesized by emulsifier-free emulsion polymerization of methyl methacrylate (MMA) at 70 °C with 2,2'-azobis(2-methylpropionamidine) dihydrochloride (AAPH) as an initiator, as reported by Stein et al. previously [14]. The resulting PMMA sphere suspension was then transferred to a glass bottle and left for several weeks at room temperature for the spheres to precipitate completely. After the evaporation of water, PMMA colloidal crystal monoliths were formed. The prepared PMMA colloidal crystal monoliths were heated at 120 °C for 15 min to induce a stronger contact between each of the spheres. The PMMA colloidal crystals in this study are composed of ca. 330 nm diameter PMMA spheres.

Synthesis of 3D Ordered Macro-/Mesoporous TiO_2 (3DOM/m TiO_2) Monoliths. Titanium tetraisopropoxide was used as the titanium source, and amphiphilic triblock copolymer Pluronic P123 ($EO_{20}PO_{70}EO_{20}$) was used as the mesopore-directing agent. The titania precursor solution was typically prepared as follows: 2.84 g titanium tetraisopropoxide (TTIP) was dissolved in 2.4 g concentrated HCl (37 wt.%) at room temperature under stirring, then a solution of 1.16 g P123 (Aldrich) dissolved in 4.0 g ethanol was added; the mixed solution was stirred at room temperature for one further hour to form a clear solution. Then the titania precursor solution was infiltrated into the void spaces of the PMMA colloidal crystal monolith by immersing the colloidal crystal in the precursor solution for 4 h. After that, the precursor-filled PMMA colloidal crystal was removed from the solution, and then aged at room temperature for 3 days and 80 °C for 1 day. Ordered macro-/mesoporous TiO_2 monolith was obtained by heating the precursor/PMMA composite in open air to 400 °C at 1 °C/min., followed by a 4-h soak to remove the PMMA template and the triblock copolymer surfactant.

Characterization. Scanning electron microscopy (SEM) images were obtained on a Hitachi S-4700 FEG scanning electron microscope. Transmission electron microscopy (TEM) and high-resolution TEM (HRTEM) were carried out on a Hitachi H800 and a JEOL JEM-3010 transmission electron microscope, respectively (both operating at 200 kV). Powder X-ray diffraction (XRD) data were collected on a Shimadzu XRD-6000 diffractometer with Cu Kα radiation (λ = 1.5418 Å). The nitrogen adsorption and desorption isotherms at 77 K were measured using a Micromeritics ASAP 2010 apparatus.

RESULTS AND DISCUSSION

Figure 1 shows a schematic view of the procedure for the synthesis of three-dimensionally ordered macro-/mesoporous (3DOM/m) TiO_2 monolith. First, a highly acidic mesoporous titania precursor solution containing titanium tetraisopropoxide (TTIP), HCl solution, triblock copolymer P123 and ethanol was infiltrated into the void spaces of the PMMA colloidal crystal monolith, followed by an aging process to form a titania gel/PMMA composite. Since TTIP is very reactive toward water, strong acid (37 wt% HCl) was added in the precursor

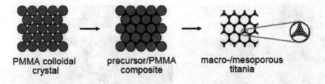

PMMA colloidal   precursor/PMMA   macro-/mesoporous
crystal   composite   titania

Figure 1. Schematic of the fabrication of ordered macro-/mesoporous TiO$_2$.

to inhibit the fast hydrolysis and condensation reaction of TTIP [15]. During the evaporation of solvent and HCl in the aging process, evaporation-induced self-assembly occurred [16], resulting in the formation of titania gel with ordered inorganic/organic mesostructure in the interstices of PMMA colloidal crystal. Then the titania gel/PMMA composite was carefully calcined in air to remove the PMMA template and the triblock copolymer surfactant, resulting in the production of 3DOM/m TiO$_2$ monolith.

Figure 2a reveals a typical SEM image of PMMA colloidal crystal monolith consisting of highly ordered close-packed PMMA spheres with 330 nm diameter. Different regions corresponding to (111) and (100) orientations can be observed. The 3DOM/m TiO$_2$ monolith (Figure 2b) produced from the PMMA colloidal crystal template exhibits a uniform porous structure that is the inverse of the original colloidal crystal. The size of the macropores is ca. 230 nm, ~ 30% smaller than the original PMMA sphere diameter due to the volume shrinkage during the calcination process.

The mesostructure and crystallization of the TiO$_2$ sample were examined by TEM and HRTEM. Figures 3a-c show the TEM images of as-prepared 3DOM/m TiO$_2$ in the (111), (110), and (100) directions, respectively. The hierarchically ordered macro-mesoporous structure can be seen in the TEM images. The HRTEM image of the macropore wall is presented in Figure 3d. It can be seen the coexistence of porous mesostructure and high crystallinity in the wall. The mesopore size is in the range of 3–5 nm, and the TiO$_2$ nanocrystallites have diameter of 7–10 nm. The lattice fringes with 0.35 nm spacing are observed in the mesopore walls, corresponding to the d-spacing between adjacent (101) crystallographic planes of anatase TiO$_2$ phase.

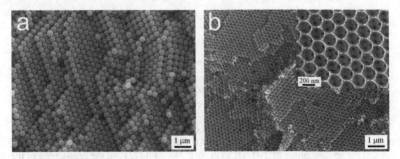

Figure 2. SEM images of (a) PMMA colloidal crystal consisting of 330 nm diameter spheres, (b) ordered macro-/mesoporous TiO$_2$ after the removal of the PMMA template, inset: higher magnification with (111) orientation.

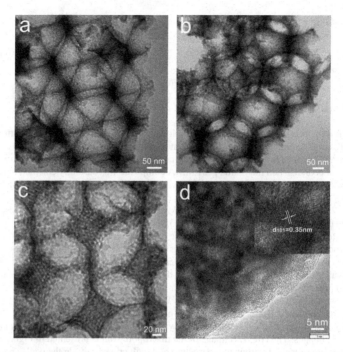

Figure 3. TEM images of ordered macro-/mesoporous TiO_2 viewed along (a) (111), (b) (110), and (c) (100) directions. (d) HRTEM image of the TiO_2 macropore wall.

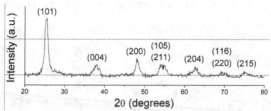

Figure 4. XRD pattern of ordered macro-/mesoporous TiO_2.

The powder XRD pattern of this sample is shown in Figure 4. The diffraction peaks in the pattern can be indexed as the (101), (004), (200), (105), (211), (204), (116), (220) and (215) reflections of the anatase phase of TiO_2 (JCPDS card, No. 21-1272) and the broadening of the peaks indicates that the walls of the mesoporous TiO_2 consist of nanocrystalline anatase, consistent with the HRTEM investigation. The average crystallite size of this sample is estimated to be about 8.5 nm from the full width at half maximum (FWHM) of the (101) diffraction peak ($2\theta = 25.4°$) using Scherrer equation, consistent with the result of the HRTEM image.

Figure 5 shows the nitrogen adsorption/desorption isotherm of as-prepared 3DOM/m TiO$_2$. The isotherm can be classified as type IV, typical for mesoporous materials according to the IUPAC classification [17]. Two hysteresis loops appear in the isotherm. The hysteresis loop at low relative pressure between 0.40 and 0.80, is of type H2, and can be ascribed to capillary condensation in mesopores generated by the surfactant P123. The other loop at high relative pressure between 0.80 and 1.0 has a type H3 shape, which may reflect the rough texture on the macropore wall surface, or incomplete filling of the precursor within the PMMA colloidal crystal [18, 19]. The Brunauer-Emmett-Teller (BET) surface area and pore volume of the 3DOM/m TiO$_2$ are 171 m^2/g and 0.402 cm^3/g, respectively. The inset in Figure 5 shows the pore size distribution calculated by Barrett-Joyner-Halenda (BJH) method using the desorption isotherm. The average pore diameter of the mesoporous TiO$_2$ is 3.7 nm. In addition, the mesopore size distribution is in the range of 3–5 nm, and such a narrow distribution implies substantial homogeneity of the mesopores for the TiO$_2$ macropore walls. The BJH pore size distribution results agree well with the TEM and HRTEM observations.

CONCLUSIONS

In summary, 3D ordered macro-/mesoporous (3DOM/m) TiO$_2$ monoliths have been successfully fabricated by a facile dual-templating approach using PMMA colloidal crystals as macropore templates and triblock copolymer Pluronic P123 as a mesopore template. The macropore walls of the prepared 3DOM/m TiO$_2$ exhibit a well-defined mesoporous structure with narrow pore size distribution, and the mesopore walls are composed of nanocrystalline anatase TiO$_2$. The material also shows a high BET surface area (171 m^2/g), and large pore volume (0.402 cm^3/g). Due to their unique porous structure and three-dimensional periodicity, the hierarchically ordered macro-/mesoporous TiO$_2$ monoliths may find numerous applications, for example, in photocatalysis, photonics, solar energy conversion, and separation.

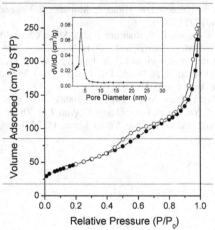

Figure 5. Nitrogen adsorption/desorption isotherm and pore size distribution curve (inset) for ordered macro-/mesoporous TiO$_2$.

ACKNOWLEDGMENTS

The authors gratefully acknowledge financial support from the National Science Foundation of China (20971012 and 20977005), National 973 Program of China (2009CB219903), National 863 Program of China (2009AA033301), and Scientific Research Foundation for the Returned Overseas Chinese Scholars, Ministry of Education of China (LX2009-04).

REFERENCES

1. X. B. Chen and S. S. Mao, Chem. Rev. **107,** 2891 (2007).
2. C. Aprile, A. Corma, and H. Garcia, Phys. Chem. Chem. Phys. **10,** 769 (2008).
3. P. D. Yang, D. Y. Zhao, D. I. Margolese, B. F. Chmelka, and G. D. Stucky, Nature **396,** 152 (1998).
4. P. C. A. Alberius, K. L. Frindell, R. C. Hayward, E. J. Kramer, G. D. Stucky, and B. F. Chmelka, Chem. Mater. **14,** 3284 (2002).
5. E. L. Crepaldi, G. J. A. A. Soler-Illia, D. Grosso, F. Cagnol, F. Ribot, and C. Sanchez, J. Am. Chem. Soc. **125,** 9770 (2003).
6. X. Y. Yang, Y. Li, A. Lemaire, J. G. Yu, and B. L. Su, Pure Appl. Chem. **81,** 2265 (2009).
7. J. G. Yu, L. J. Zhang, B. Cheng, and Y. R. Su, J. Phys. Chem. C **111,** 10582 (2007).
8. Y. N. Fu, Z. G. Jin, W. J. Xue, and Z. P. Ge, J. Am. Ceram. Soc. **91,** 2676 (2008).
9. J. Q. Zhao, P. Wan, J. Xiang, T. Tong, L. Dong, Z. N. Gao, X. Y. Shen, and H. Tong, Micropor. Mesopor. Mater. **138,** 200 (2011).
10. J. I. L. Chen, G. von Freymann, S. Y. Choi, V. Kitaev, and G. A. Ozin, Adv. Mater. **18,** 1915 (2006).
11. L. I. Halaoui, N. M. Abrams, and T. E. Mallouk, J. Phys. Chem. B **109,** 6334 (2005).
12. X. C. Wang, J. C. Yu, C. M. Ho, Y. D. Hou, and X. Z. Fu, Langmuir **21,** 2552 (2005).
13. J. Konishi, K. Fujita, K. Nakanishi, K. Hirao, K. Morisato, S. Miyazaki, and M. Ohira, J. Chromatogr. A **1216,** 7375 (2009).
14. R. C. Schroden, M. Al-Daous, C. F. Blanford, and A. Stein, Chem. Mater. **14,** 3305 (2002).
15. K. X. Wang, B. D. Yao, M. A. Morris, and J. D. Holmes, Chem. Mater. **17,** 4825 (2005).
16. C. J. Brinker, Y. F. Lu, A. Sellinger, and H. Y. Fan, Adv. Mater. **11,** 579 (1999).
17. K. S. W. Sing, D. H. Everett, R. A. W. Haul, L. Moscou, R. A. Pierotti, J. Rouquerol, and T. Siemieniewska, Pure Appl. Chem. **57,** 603 (1985).
18. Z. Y. Wang and A. Stein, Chem. Mater. **20,** 1029 (2008).
19. Y. H. Deng, C. Liu, T. Yu, F. Liu, F. Q. Zhang, Y. Wan, L. J. Zhang, C. C. Wang, B. Tu, P. A. Webley, H. T. Wang, and D. Y. Zhao, Chem. Mater. **19,** 3271 (2007).

Mater. Res. Soc. Symp. Proc. Vol. 1352 © 2011 Materials Research Society
DOI: 10.1557/opl.2011.1010

Growth of TiO₂ thin films on Si(001) and SiO₂ by reactive high power impulse magnetron sputtering

F. Magnus[1], B. Agnarsson[1], A. S. Ingason[1,2], K. Leosson[1], S. Olafsson[1], and J. T. Gudmundsson[1,3]

[1]Science Institute, University of Iceland, Dunhaga 3, IS-107, Reykjavik, Iceland
[2]Thin Film Physics, Department of Physics (IFM), Linköping University, Linköping SE-581 83, Sweden
[3]UM-SJTU Joint Institute, Shanghai Jiao Tong University, 800 Dong Chuan Road, Shanghai, 200240, China

ABSTRACT

Thin TiO₂ films were grown on Si(001) and SiO₂ substrates by reactive dc magnetron sputtering (dcMS) and high power impulse magnetron sputtering (HiPIMS) at temperatures ranging from 300 to 700 °C. Both dcMS and HiPIMS produce polycrystalline rutile TiO₂ grains, embedded in an amorphous matrix, despite no postannealing taking place. HiPIMS results in significantly larger grains, approaching 50% of the film thickness at 700 °C. In addition, the surface roughness of HiPIMS-grown films is below 1 nm rms in the temperature range 300–500 °C which is an order of magnitude lower than that of dcMS-grown films. The results show that smooth, rutile TiO₂ films can be obtained by HiPIMS at relatively low growth temperatures, without postannealing.

INTRODUCTION

TiO₂ finds application in a variety of electrical and optical devices due to its high refractive index and good thermal stability. In bulk form, TiO₂ is known to exist in three crystalline structures; two tetragonal structures, the anatase phase and the rutile phase; and an orthorhombic structure, the brookite phase. In its thin film form, only anatase and rutile are observed [1, 2]. When TiO₂ is grown at low temperature it tends to be amorphous [2]. At higher temperatures (up to 600 °C) the anatase phase is favored whereas above 600 °C the rutile phase starts to appear. By postannealing above 800 °C rutile phase dominated films are obtained but this transformation also strongly depends on the deposition temperature, the ion energy and the energy flux during growth [1-5]. In its rutile phase, the transparency and high refractive index make TiO₂ attractive for the glass coating industry, where it is used in low emissivity and antireflective coatings [1]. In microelectronic applications, the high dielectric constant ($\kappa = 80$) is also of interest [1, 6]. TiO₂ coatings in the anatase phase are used as photocatalysts, as self cleaning and antibacterial surfaces [7, 8], and for coating of biomedical surfaces [9]. TiO₂ is also hard and chemically resistant which can be useful in many applications.

High power impulse magnetron sputtering (HiPIMS) is a novel ionized physical vapor deposition (IPVD) technique that has shown great promise in materials processing [10]. By pulsing the target with short unipolar voltage pulses at a low frequency and low duty cycle, a high electron density is achieved which leads to a high ionization fraction of the sputtered vapor [11]. Davis et al. [12] have demonstrated growth of TiO₂ thin films by reactive HiPIMS from a

Ti target. The resulting films exhibited a slightly higher refractive index but also lower deposition rates compared to dc magnetron sputtered (dcMS) films given the same average target power. Konstantinidis et al. [13] showed that dense rutile films can be produced by HiPIMS at growth temperatures as low as room temperature when the films are grown reactively from a metallic target on a stainless steel substrate. However, when grown on a glass substrate the films were in the anatase phase. Sarakinos et al. [14] applied HiPIMS to grow TiO_x ($x > 1.8$) films reactively from a ceramic $TiO_{1.8}$ target. They produced films with lower surface roughness, higher densities and higher refractive indices in comparison to the films grown by dcMS. Finally, Alami et al. [15] have shown that rutile films, with TiO_2 crystals embedded in an amorphous matrix, can be obtained on Si(001) substrates without substrate heating, by HiPIMS.

The aim of this study is to compare thin TiO_2 films, grown by reactive HiPIMS and dcMS at substrate temperatures ranging from 300 to 700 °C, which have not received the standard postannealing treatment. The morphology of TiO_2 films grown on Si(001) and amorphous SiO_2 substrates is compared. Crystallinity and phase composition is examined *ex-situ* by grazing incidence X-ray diffraction (GIXRD) measurements and low-angle X-ray reflectivity (XRR) measurements are performed to determine the film thickness, density and roughness.

EXPERIMENT

The TiO_2 thin films were grown in a custom built magnetron sputtering chamber with a base pressure of 1×10^{-6} Pa. The sputtering gas was argon of 99.999% purity mixed with oxygen gas of 99.999% purity. The argon flow rate was $q_{Ar} = 37$ sccm and the oxygen flow rate $q_{O2} = 1.2$ sccm and a throttle valve was used to set a total growth pressure of 0.7 Pa. The substrates used were Si(001) with an approximately 2 nm thick native oxide layer and thermally oxidized Si(001) with an oxide thickness of roughly 500 nm (from now on referred to as the SiO_2 substrate). The substrate temperature was controlled during growth with a 2 inch rectangular bore-nitride plate heater, separated from the substrate holder by a 10 mm gap. A Ti target was used, 75 mm in diameter and of 99.995% purity. The target-substrate distance was approximately 200 mm and the substrate surface was at an angle of approximately 45° to the target.

Film deposition was carried out by both dcMS and HiPIMS. dcMS was carried out in constant power mode at 140 W using an Advanced Energy MDX500 power supply. For HiPIMS the power was supplied by a SPIK1000A pulse unit (Melec GmbH) operating in the unipolar negative mode at a constant voltage, which in turn was charged by a dc power supply (ADL GS30). The discharge current and voltage was monitored using a combined current transformer and voltage divider unit (Melec GmbH) and the data were recorded with a digital storage oscilloscope (Agilent 54624A). The cathode voltage waveform was square and approximately 530 V. The pulse length was 200 µs and the pulse repetition frequency was 75 Hz. The peak power density was 0.70 ± 0.05 kW/cm^2 and the power averaged over one period was 350 ± 20 W. During both dcMS and HiPIMS the discharge was kept in metal mode and the average current was similar, or in the range 550–650 mA. The growth time was 30 minutes.

Phase identification and grain size measurements were carried out by grazing incidence X-ray diffractometry (GIXRD) using a Philips X'pert diffractometer (Cu K_a, wavelength 0.15406 nm) mounted with a hybrid monochromator/mirror on the incident side and a 0.27° collimator on the diffracted side. A line focus was used with a beam width of approximately 1 mm. The GIXRD scans were carried out with the incident beam at $\theta < 1°$. The film thickness, density and roughness were determined by low-angle X-ray reflectivity (XRR) measurements with an angular resolution of 0.005°. The low incident angle means that these measurements probe almost the entire film area.

RESULTS AND DISCUSSION

All the films grown by both HiPIMS and dcMS are polycrystalline as seen by the presence of peaks in GIXRD scans. However, a large sloping background indicates that part of the films is in an amorphous phase. Examples of GIXRD scans of the films grown by HiPIMS are shown in Figure 1. In all films except one, the observed peaks correspond to the rutile phase and the rutile peaks become more prominent with increasing growth temperature. The exception is the film grown by HiPIMS on SiO_2 at 300 °C where a small anatase (101) peak is present. The absence of the anatase phase in most of our films is somewhat surprising, as the growth of anatase is usually strongly favored with respect to rutile at temperatures below 600 °C [1].

Despite the presence of the native oxide layer, the Si (001) substrates clearly promote the nucleation of rutile grains. This can also be seen in Figure 2 which shows the grain size calculated from the FWHM of the rutile (110) peaks. The grain size is normalized by the thickness of the films, as the dcMS-grown films are twice as thick as the HiPIMS-grown films. It should be noted that the growth time is the same in both cases, eliminating grain growth through the motion of grain boundaries as a factor to take into account [16]. HiPIMS results in significantly larger grains than dcMS and the grain size for the films grown by HiPIMS at 700 °C is approaching half the film thickness. This is a result of the enhanced energetic ion bombardment during HiPIMS. The ion flux to the substrate is composed of moderate energy positive Ar and Ti ions and high energy O⁻ ions, and their impact with the growing film promotes the formation of rutile grains [15].

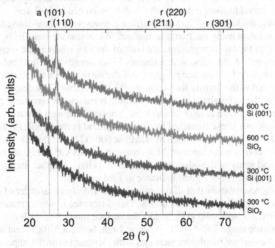

Figure 1. Grazing incidence x-ray diffraction measurements of HiPIMS-grown TiO_2 films. The curves have been shifted for clarity. The growth temperature and substrate is shown on the right and the peaks are labeled by "a" for anatase and "r" for rutile.

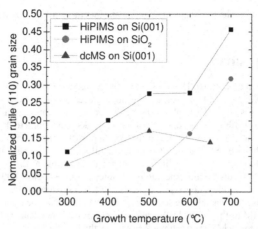

Figure 2. Grain size of the rutile (110) grains, normalized by the film thickness, as a function of substrate growth temperature.

The most striking difference between the films grown by HiPIMS and dcMS is revealed by XRR measurements. A comparison between the specular XRR from dcMS-grown and HiPIMS-grown films at three different growth temperatures is shown in Figure 3. The drop in reflected intensity is mostly determined by the surface roughness of the films. It is immediately obvious from Figure 3 that the films grown by dcMS are much rougher than the films grown by HiPIMS, although this could to some extent be due to the difference in thickness.

The film density, thickness and surface/interface roughness can be extracted from the XRR scans by fitting the data using the Parrat formalism. We obtain the best fits by using a two-layer model where one of the layers represents the bulk of the film whereas the second layer is a surface layer, approximately 5 nm thick, with a slightly lower density than the "bulk" layer. The density of the "bulk" layer and the surface roughness of the surface layer are shown in Figure 4. We observe no clear trend in the density data with growth temperature. The density values are lower than expected for the rutile phase (4.2 g/cm^3), which is somewhat surprising given the large grain size in the films grown at high temperature. This can be a sign that the films are porous or that the amorphous phase (revealed by the background in the GIXRD scans) has a low density. The highest density value is obtained by dcMS at 500 °C but as this film does not produce the largest crystallites it is clear that the increase in density can not be attributed to rutile grains. Since both Ti and many of the other oxide phases of Ti have a higher density than the TiO$_2$ phases the density alone is not a good indicator of film quality.

The XRR fitting results show that films grown by HiPIMS have an order of magnitude lower surface roughness than films grown by dcMS. This difference is much greater than can be accounted for by statistical roughening [17]. In fact the surface roughness for films grown by HiPIMS in the temperature range 300–500 °C is below 1 nm, both for Si(001) and SiO$_2$ substrates. However, the surface roughness increases with increased growth temperature. The smoothening of film surfaces during HiPIMS has also been attributed to energetic ion bombardment [18].

Thickness values obtained from the XRR measurements of dcMS-grown films are somewhat unreliable due to the high roughness. However, we estimate that thickness of the films grown by dcMS is 100–120 nm whereas the thickness of the films grown by HiPIMS is 50–60 nm. As the growth time is the same for both methods and the average current is similar we can deduce that the growth rate during HiPIMS is approximately halved compared to dcMS.

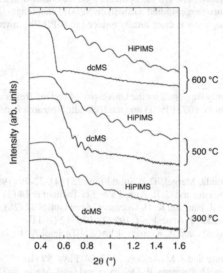

Figure 3. XRR measurements of TiO₂ films grown by HiPIMS and dcMS at the three different growth temperatures shown on the right. The curves have been shifted for ease of comparison.

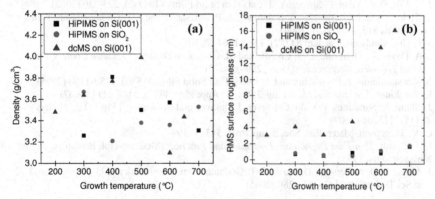

Figure 4. The density (a) and rms surface roughness (b) of the TiO₂ films as a function of substrate growth temperature, as determined by fitting the XRR data by the Parrat formalism.

CONCLUSIONS

Thin TiO$_2$ films were grown on Si(001) and SiO$_2$ substrates at various growth temperatures by dcMS and HiPIMS. No postannealing was employed. Both methods produce rutile phase films with grain size increasing with growth temperature, although the rutile grains coexist with an amorphous phase. The HiPIMS process produces films with a higher degree of crystallinity in the temperature range studied. Growth on Si(001) encourages the formation of rutile grains. The surface roughness is dramatically reduced in HiPIMS-grown films compared to dcMS films.

ACKNOWLEDGMENTS

This work was partially supported by the University of Iceland Research Fund, the Icelandic Research Fund grant no 072105003 and the Icelandic Research Fund Excellence Grant no 100019011.

REFERENCES

1. P. Lobl, M. Huppertz and D. Mergel, Thin Solid Films **251** (1), 72-79 (1994).
2. M. D. Wiggins, M. C. Nelson and C. R. Aita, J Vac Sci Technol A **14** (3), 772-776 (1996).
3. K. Balasubramanian, X. F. Han and K. H. Guenther, Appl Optics **32** (28), 5594-5600 (1993).
4. S. Mraz and J. M. Schneider, J. Appl. Phys. **109** (2), 023512 (2011).
5. D. Wicaksana, A. Kobayashi and A. Kinbara, J Vac Sci Technol A **10** (4), 1479-1482 (1992).
6. G. D. Wilk, R. M. Wallace and J. M. Anthony, J. Appl. Phys. **89** (10), 5243-5275 (2001).
7. K. Eufinger, D. Poelman, H. Poelman, R. De Gryse and G. B. Marin, in *Thin Solid Films: Process and Applications*, edited by N. C. Nam (2008).
8. I. Salem, Catal Rev **45** (2), 205-296 (2003).
9. J. X. Liu, D. Z. Yang, F. Shi and Y. J. Cai, Thin Solid Films **429** (1-2), 225-230 (2003).
10. U. Helmersson, M. Lattemann, J. Bohlmark, A. P. Ehiasarian and J. T. Gudmundsson, Thin Solid Films **513** (1-2), 1-24 (2006).
11. J. T. Gudmundsson, Vacuum **84** (12), 1360-1364 (2010).
12. J. A. Davis, W. D. Sproul, D. J. Christie and M. Geisler, in *Society of Vacuum Coaters 47th Annual Technical Conference* (Dallas, 2004), pp. 215.
13. S. Konstantinidis, J. P. Dauchot and A. Hecq, Thin Solid Films **515** (3), 1182-1186 (2006).
14. K. Sarakinos, J. Alami and M. Wuttig, J Phys D Appl Phys **40** (7), 2108-2114 (2007).
15. J. Alami, K. Sarakinos, F. Uslu, C. Klever, J. Dukwen and M. Wuttig, J Phys D Appl Phys **42** (11), 115204 (2009).
16. C. V. Thompson, Mater. Res. Soc. Symp. Proc. **343**, 3 (1994).
17. D. L. Smith, *Thin-Film Deposition: Principles and Practice*. (McGraw-Hill, Boston, Massachusetts, 1995).
18. J. Alami, P. O. A. Persson, D. Music, J. T. Gudmundsson, J. Bohmark and U. Helmersson, J Vac Sci Technol A **23** (2), 278-280 (2005).

Properties

Mater. Res. Soc. Symp. Proc. Vol. 1352 © 2011 Materials Research Society
DOI: 10.1557/opl.2011.1052

Dynamics of Water Confined on the Surface of Titania and Cassiterite Nanoparticles

Nancy L. Ross,[1] Elinor C. Spencer,[1] Andrey A. Levchenko,[2] Alexander I. Kolesnikov,[3] Douglas L. Abernathy,[3] Juliana Boerio-Goates,[4] Brian F. Woodfield,[4] Alexandra Navrotsky,[5] Guangshe Li,[6] Wei Wang,[7] David J. Wesolowski[7]

[1]Dept. of Geosciences, Virginia Tech, Blacksburg, Virginia 24061, U.S.A.
[2]Setaram Inc., 8430 Central Ave., Suite C and 3D, Newark, California 94560
[3]Neutron Scattering Sciences Division, Oak Ridge National Laboratory, PO BOX 2008, Oak Ridge, Tennessee 37831, U.S.A.
[4]Dept. of Chemistry and Biochemistry, Brigham Young University, Provo, Utah 84602
[5]Peter A. Rock Thermochemistry Laboratory and NEAT ORU, University of California at Davis, Davis, California 95616, U.S.A.
[6]State Key Lab of Structural Chemistry, Fujian Institute of Research on the Structure of Matter, Chinese Academy of Science, Fuzhou 350002, P. R. China.
[7]Chemical Science Division, Oak Ridge National Laboratory, PO BOX 2008, Oak Ridge, Tennessee 37831, U.S.A.

ABSTRACT

We present low-temperature inelastic neutron scattering spectra collected on two metal oxide nanoparticle systems, isostructural TiO_2 rutile and SnO_2 cassiterite, between 0-550 meV. Data were collected on samples with varying levels of water coverage, and in the case of SnO_2, particles of different sizes. This study provides a comprehensive understanding of the structure and dynamics of the water confined on the surface of these particles. The translational movement of water confined on the surface of these nanoparticles is suppressed relative to that in ice-Ih and water molecules on the surface of rutile nanoparticles are more strongly restrained that molecules residing on the surface of cassiterite nanoparticles. The INS spectra also indicate that the hydrogen bond network within the hydration layers on rutile is more perturbed than for water on cassiterite. This result is indicative of stronger water-surface interactions between water on the rutile nanoparticles than for water confined on the surface of cassiterite nanoparticles. These differences are consistent with the recently reported differences in the surface energy of these two nanoparticle systems.

INTRODUCTION

Metal oxide nanoparticles have potential application in a number of important fields such as catalysis, environmental remediation, energy conversion and sensor technology.[1-4] The existence of hydration layers on their surface can play a critical role in stabilizing the nanoparticle[5-7], yet the dynamics and physical properties of water are altered with respect to bulk by confining it to nanoscale domains, surfaces, or interfaces.[8,9] Thus the development of stable nanoparticles for practical uses requires that the properties of these hydration layers be assessed.

The large incoherent neutron scattering cross-section of hydrogen relative to transition metals and the absence of selection rules ensures that inelastic neutron scattering (INS) techniques are ideal for probing the dynamics of the nanoparticle hydration layers without significant interference from the underlying metal oxide lattice.[10] Herein we compare INS spectra collected on hydrated TiO_2 rutile[11] (1) with SnO_2 cassiterite (2) nanoparticles with different levels of water coverage and various particle sizes. TiO_2 rutile and SnO_2 cassiterite are isostructural and permit direct comparison of their hydrated forms without concern about differences in the basic structure and symmetry. Additionally, the surface energies of TiO_2 and SnO_2 nanoparticles are 2.22 ± 0.07 and 1.72 ± 0.01 Jm^{-1}, respectively, which are sufficiently different to accentuate any variations in the INS spectra of these materials.[12] Indeed, static planewave density functional theory (PW-DFT) calculations predict significant differences in the nature of water adsorbed on TiO_2 rutile and SnO_2 cassiterite (1 1 0) crystal surfaces, with water dissociation on the SnO_2 surface being energetically more favorable than on the TiO_2 surface.[13] We therefore expect the ratio of molecular to dissociative water to be different on the surfaces of TiO_2 and SnO_2 nanoparticles leading to alterations in the dynamic behavior of the confined surface water and appreciable differences in their INS spectra.

EXPERIMENT

The preparation of TiO_2 is documented elsewhere.[11] These nanoparticles are known to have a rod-like morphology and a uniform size distribution of 35 nm (length) by 7 nm (diameter). The as-synthesized sample will be denoted as **1A** and has water coverage of 7.6 wt.% and surface area of 96 m^2g^{-1}. The chemical formula of **1A** is therefore $TiO_2 \cdot 0.37H_2O$. This corresponds to 4.760×10^{-5} moles of water *per* m^2 of rutile surface (29 molecules of water *per* nm^2). A sample of **1A** was partially dehydrated by being heated at 100°C under vacuum to give sample **1B**. From measurements of water loss during the dehydration procedure the water coverage for **1B** is 2.817×10^{-5} moles of water *per* m^2 of rutile surface (4.7 wt.%, 17 molecules of water *per* nm^2). Consequently, the chemical formula for **1B** is $TiO_2 \cdot 0.22H_2O$.

Nanoparticles of SnO_2 were synthesised using the methodology reported by Mamontov et al.[14] A 0.1 M solution of $SnCl_4$ was placed in a Teflon-lined steel autoclave and heated at 423 K for 16 h. The resultant SnO_2 colloid was purified by dialysis, dispersed in pure water by sonication, and then kept at 423 K for an additional 16 h before being freeze-dried. The dry nanopowder was hydrated prior to the experiment by allowing the powder to equilibrate with laboratory air (~296 K and 80% relative humidity) for ~24 h. Nanoparticles prepared in this manner are known to have a truncated rod-like morphology with an average diameter of 4 nm and average length of 7 nm (corresponding to an average aspect ratio of 1:1.75) with a surface area is 155.5 m^2g^{-1}. The as-synthesized sample will be denoted as **2A**. From water loss measurements the water coverage for **2A** is ~11 wt.%. This is equivalent to 4.412×10^{-5} moles of water *per* m^2 of cassiterite surface (27 molecules of water *per* nm^2). The chemical formula for **2A** is $SnO_2 \cdot 1.03H_2O$. Sample **2A** was dehydrated in a similar manner to **1A** to give a dehydrated sample that we will denote as **2B**. The INS spectra of **2B** (see Results and Discussion section) indicated that no molecular water remains on the surface of the particles implying that the hydration layers on **2A** were fully stripped from the particles during the dehydration procedure leaving only hydroxyl groups chemically bound to the nanoparticle surface.

A sample of 2 nm SnO_2 nanoparticles (**2C**) was synthesised by a previously published synthetic method with some modifications.[15] Anhydrous $SnCl_4$ was mixed with excess solid NaOH and then sufficient H_2O was added to create a slurry, and then the mixture was ground for 10 min with a pestle and mortar. The resulting solid mixture was rinsed thoroughly with water until Cl^- ions could no longer be detected by the addition of a concentrated solution of $Ag+$ ions. The final product was confirmed to be phase pure. A 6 nm SnO_2 sample (**2D**) was synthesized by calcining a portion of the 2 nm SnO_2 sample at 500 C for 2 h. Nanoparticles of SnO_2 20nm in size SnO_2 (**2E**) were purchased from Sigma-Aldrich and used after exposure to the open atmosphere (294 K, ~70% relative humidity) for 30h. The water contents of samples **2C–E** were determined by thermogravimetric analyses (TGA) to be 8.93 wt.% (**2C**), 1.12 wt.% (**2D**) and 1.09 wt.% (**2E**); therefore the chemical formulae for these samples are $SnO_2{\cdot}0.82H_2O$ (**2C**), $SnO_2{\cdot}0.095H_2O$ (**2D**), and $SnO_2{\cdot}0.092H_2O$ (**2E**).

INS data for samples **1A**, **1B**, **2A**, and **2B** were collected on the HRMECS spectrometer at the Intense Pulsed Neutron Source, which is now shut-down (Argonne National Laboratory, Argonne, IL),[16,17] and INS spectra for samples **2C**, **2D** and **2E** were measured with the ARCS spectrometer at the Spallation Neutron Source (Oak Ridge National Laboratory, Oak Ridge, TN).[18] Both spectrometers are direct geometry time-of-flight instruments for which the energy of the incident neutrons (E_i) can be selected by rotating Fermi choppers. The energy resolutions of these spectrometers are very similar, $\Delta E/E_i \approx 2–5\%$. The neutron flux on the samples placed on ARCS was more than 100 times higher than for the samples analyzed on HRMECS. INS spectra were collected on **1A** (31.20 g), **1B** (29.83 g), **2A** (50.15 g), and **2B** (73.79 g) at 4-10 K, and on **2C** (4.39 g), **2D** (6.17 g) and **2E** (8.04 g) at 18 K. Data were collected on HRMECS with E_i set to 50, 140 and 600 meV, and on ARCS with E_i equal to 60, 160 and 600 meV, therefore the total energy range covered was $0 < E <550$ meV. The momentum transfer range (Q) was 0.5-11.0 Å^{-1} and is dependent on E_i, energy transfer, and the scattering angle. Background measurements were collected with an empty sample can under similar conditions to those employed for the sample data collections. These background data were subsequently subtracted from the sample data. The measured INS data were initially transformed into dynamical structure factors, $S(Q,E)$, and then converted to the generalized vibrational density of states, $G(E)$, according to the standard expression

$$G(E) = \frac{S(Q,E)E \exp(<u_H^2>Q^2)}{Q^2[n(E,T)+1]} \tag{1}$$

where $n(E,T)$ is the population Bose factor, and $<u_H^2>$ is the mean squared displacement of the hydrogen atoms of water.

DISCUSSION

INS spectra for **1** and **2** are shown in Figure 1 along with the INS spectrum of ice-Ih for comparison.[19] If we assume that the water molecules conform to a rigid body approximation, then the translational motions along the three crystallographic axes are expected to occur with the same magnitude in each direction for all atoms of a water molecule, i.e. their phonon eigenvectors are the same. The librational modes of water correspond to the rotational motion of the 'rigid' molecules about the three axes of rotation that pass through, or very close to, the oxygen atoms of the molecules (consequently, the involvement of the oxygen atoms in the librational motion of the molecules is negligible). The energies of the translational motions of the water molecules in ice are inversely proportional to the square root of the molecular mass. The energies of the librational modes are inversely proportional to the square root of the non-isotropic moments of inertia of the molecules. Therefore comparison of INS spectra of hydrogenated and deuterated ices allows for the identification of the translational and librational

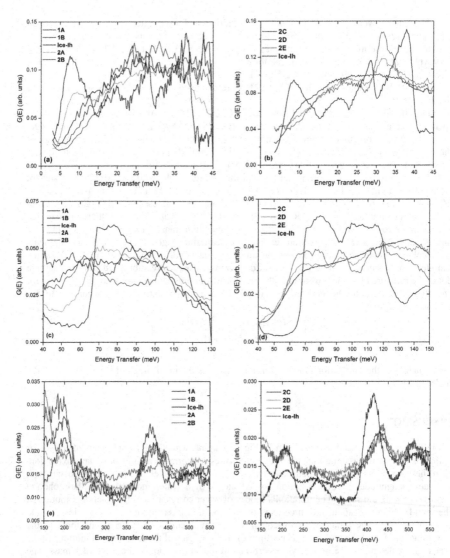

Figure 1. INS spectra [$G(E)$] for nanoparticles of **1** and **2** and bulk ice-Ih obtained with HRMECS: (a) E_i = 50 meV, (c) E_i = 140 meV, (e) E_i = 600 meV; and ARCS: (b) E_i = 60 meV, (d) E_i = 160 meV, (f) E_i = 600 meV. Data for **1** are taken from ref. 11.

50

bands in the spectra. From such studies we know that the translational and librational modes for water are observed in the energy ranges 0-40 meV and 50-120 meV, respectively.[20]

The center of gravities (CoG) of the translational and librational bands in the ice-Ih and nanoparticle spectra were calculated with the following equation:

$$CoG = \frac{\int_a^b E \cdot g(E)^p \, dE}{\int_a^b g(E)^p \, dE} \qquad (2)$$

where a = lower limit of the energy range; b = upper limit of the energy range; P = weighting parameter. The limits of integration are $a = 4$ meV and $b = 40$ meV for the translation bands (the region 0-3.5 meV is obscured by the elastic signal), and $a = 50$ meV and $b = 120$ meV for the librational bands. In both case, the parameter P was set such that the weighting is against the absolute spectrum ($P = 1$). The results from these calculations are reported in Table 1.

Table 1. CoG values for the translational and librational bands of **1**, **2**, and ice-Ih.

	CoG for translational band (meV)	CoG for librational band (meV)
TiO₂		
1A	24.6	83.4
1B	25.1	83.1
SnO₂		
2A	23.8	85.9
2B	26.2	85.8
2C	23.9	88.9
2D	24.4	87.9
2E	24.8	86.7
Ice-Ihᵃ	22.8 (24.5)	88.9 (92.51)

(a) Determined from data collected on HRMECS, values in brackets calculated from ARCS data.

The CoG values for **2C–E** given in Table 1 are slightly larger than those calculated from INS spectra, for the same samples, collected on TOSCA (ISIS Facility, Rutherford-Appleton Laboratory, UK).[21] Nonetheless, the data in Table 1 shows that there is no significant variation in the CoG values for the translational bands of these samples, which is consistent with the TOSCA data. The variation in the CoG values may be due to differences in the resolution of these instruments, TOSCA has a resolution of $\Delta E/E_i \approx 2\%$ over the 0–200 meV energy range, whereas both HRMECS and ARCS have resolutions of $\Delta E/E_i \approx 2$–5%. Furthermore, there is a difference in the momentum transfer (Q) range covered by the INS experiments performed on these instruments; the Q range is smaller in the case of the TOSCA data.

Translational modes of confined H₂O

The translational band observed in the spectrum of dehydrated SnO₂ (**2B**) is due to the correlated motion of the chemisorbed surface hydroxyl groups with the acoustic vibrations of the SnO₂ lattice, whereas the translational bands in the spectra for **1A**, **1B**, **2A** and **2C–E** are primarily due to the translational modes of the physisorbed molecular water on the particle surface that occur at lower energy than the metal oxide lattice vibrations, which is reflected in the reduced CoG values for the translation bands in the spectra of these samples relative to that of **2B**. The center of gravity (CoG) values for the translational bands in the spectra for **1A**, **1B**, **2A** and **2B** collected on HRMECS are at higher energy values than those in ice-Ih (Table 1). This energy redistribution is suggestive of suppression of the low energy translational modes of the confined water molecules due to strong interactions between the water and the surface of the metal oxide nanoparticles.

In Figures 1a and 1b the data are scaled so that the total area of the translational areas of the spectra are equal to three, corresponding to the three degrees of translational freedom for water molecules. By scaling the translational regions in this way it is possible to compare directly the features observed in the spectra. It is interesting to note that the first acoustic peak, which is at 7 meV in the ice-Ih spectrum, is slightly shifted to high energy in the spectrum of **2A** (ca. 8 meV) and is strongly suppressed in the spectra of **1A** and **1B**. Samples **1A** and **2A** have a similar number of water molecules per unit area of 29 and 27 molecules per nm^2, respectively, and one might therefore expect the physisorbed hydration layers on these two isostructural particles to be similar. However, the reduction in the intensity of the acoustic mode in the spectra of **1A** relative to **2A** is indicative of the water on TiO$_2$ nanoparticles being more tightly bound and thus exhibiting restricted translational motion with respect to water confined on the surface of the SnO$_2$ particles. This is likely to be due to TiO$_2$ nanoparticles possessing a greater surface energy than SnO$_2$ particles.[12] The greater hindrance of the translational diffusion of water confined on the surface of TiO$_2$ than in SnO$_2$ has also been observed in quasielastic neutron scattering (QENS) measurements of hydrated TiO$_2$ rutile and SnO$_2$ cassiterite nanoparticles.[14]

We note that the acoustic mode at 7 meV is much suppressed in the INS spectrum of **2C** and not visible in **2D** and **2E** (Fig. 1b). However, this mode is observed in the INS spectra of these materials collected on TOSCA[21]. This may be due to the greater resolution, especially at low energy transfer, of TOSCA relative to ARCS and due to differences in the Q ranges of the data, as mentioned above.

When compared to the translational bands in the ice-Ih spectrum, the corresponding bands in the spectra of the relatively highly hydrated samples **1A**, **1B**, **2A**, and **2C** lack observable structure. This implies that the energy of translational motions of the water confined on the surface of these nanoparticles is isotropic i.e. not dependent on the direction of motion relative to the crystallographic axes of the metal oxide lattice. The peak at ca. 35 meV in the spectra of **2D** and **2E** is probably due to some optical modes of the SnO$_2$ lattices, and/or riding modes of water on the surface of the particles. This peak is absent in the spectrum of **2C** presumably because the higher hydration level of this sample relative to **2D** and **2E** results in the translational signals of the physisorbed water obscuring the signals arising from these modes. In addition, due to the small particle size of sample **2C** (2 nm) the bands in the INS spectrum of this sample that correspond to the phonon modes of the SnO$_2$ lattice will be broad and lack structure due to the absence of long range order, and this will be reflected in the peaks associated with the riding modes of the surface water.

There is a small difference between the translational band CoG values for **1A** and **1B** (TiO$_2$ nanoparticles of the same size) and a large difference between the values for **2A** and **2B** (SnO$_2$ nanoparticles of the same size) that may suggest the translational band shifts to slightly higher energy with decreasing hydration level. However, it is possible that the large difference observed between the values of **2A** and **2B** is a result of the difference in the water species on the particle surface. **2A** contains a large amount of molecular water, whereas only hydroxyl groups are present on the surface of the particles comprising sample **2B**. Moreover, the similarity between the CoG values of **2D** and **2E** (samples with similar water contents) reported in Table 1 suggests that there is no significant shift in the translational energy of the surface water with increasing nanoparticle size, at least for SnO_2 particles over the 6–20 nm range.

Librational modes of confined H_2O

As stated above the librational modes for water occur within the 50-120 meV energy range of the INS spectra; the data covering this energy range are shown in Figures 1c and 1d. To permit direct comparison of the features observed in these plots the spectra have been scaled such that the total area under the librational regions equals three, corresponding to the three degrees of librational freedom for the water molecules. From inspection of these figures it is apparent that there are essentially no vibrational modes for ice-Ih in the 40-65 meV energy range, and the librational band is limited to the 65-120 meV energy region. In contrast, the librational bands for nanoparticle samples extend over the 40-120 meV energy range. The presence of the additional vibrational modes in the 40-65 meV range in the spectra for **1A** and **1B** can be ascribed to correlation of the vibrations of the hydrogen atoms of the physisorbed water and surface-bound hydroxyl groups with the intense optical modes of the TiO_2 lattice that exist within the 35-65 meV energy range. A similar phenomenon was observed in the INS spectra for hydrated anatase-TiO_2 nanoparticles.[22] An analogous rationale can be applied to explain the additional librational modes observed in the INS spectra of the hydrated SnO_2 nanoparticle samples **2A**, and **2C–E**. The relative positions of the librational bands in the spectra of **1** and **2** are discussed below.

Samples **1A**, **2A** and **2C** have very high water contents (7.6, ca. 11, and 8.93 wt.%, respectively), and the librational regions of these spectra are featureless. Conversely, sample **1B** has a lower water content (4.7 wt.%) than **1A** and a broad peak is observable at 78-110 meV that is possibly due to overlapping peaks arising from the wagging and twisting librational motions of the water molecules. The spectra for the poorly hydrated samples **2B**, **2D** and **2E** are dominated by signals arising from vibrations of the hydrogen atoms associated with the hydroxyl groups that are chemically bound to the surface of the SnO_2 cassiterite particles. The peaks at 62 meV and 110 meV in the spectrum of **2B** are probably associated with the bending in different directions of the O-H of the surface hydroxyl groups. These peaks are hidden in the spectra for **2A** by the librational band arising from the motion of the physisorbed molecular water, and are seen as broad features at 55-80 meV and as bands commencing at ca. 105 meV, in the spectra of **2D** and **2E**. The broadness of these bands relative to the corresponding peaks in the spectrum of **2B** is due to the interaction of the hydroxyl groups with the physisorbed water that is present in these samples, but not in sample **2B**; it is the librational motion of these water molecules that is responsible for the peaks at 93 meV in the spectra of **2D** and **2E**, and which is absent from the spectrum of **2B**.

The librational motion of the confined water is strongly influenced by the hydrogen bond environment of the water molecules. The CoG values for librational bands associated with water confined on the surface of the metal oxide nanoparticle are lower than the corresponding value for ice-Ih suggesting that the water molecules of the hydration layers experience a more disrupted and softer hydrogen bond environment relative to the molecules in ice-Ih. Furthermore, there is a clear difference in the librational band CoG values for samples **1** and **2**. The reduced

CoG value for **1** relative to **2** suggests that the water molecules on the surface of TiO_2 rutile particles are subjected to a more distorted hydrogen bond environment than the water confined on the SnO_2 cassiterite nanoparticles. This is no doubt due to the increased strength of the water-oxide interactions in the TiO_2 nanoparticle systems with respect to the SnO_2 samples. Furthermore, the librational band CoG values for **2D** and **2E** (samples with similar water contents) are slightly different, suggesting that the particle size may have a small influence the energy of the librational motion of the water confined to the surface of the particles.

Confined H₂O modes between 150-550 meV

Table 2. Details for key peaks the 150–550 meV range of the INS spectra of **1** and **2**.[a]

	$\delta(H_2O)$ (meV)	$\upsilon_1(H–O)$ (meV)
1A	195 (40)	417 (72)
1B	194 (51)	417 (71)
2A	206 (35)	411 (49)
2B	–	441 (25)
2C	210 (33)	427 (60)
2D	209 (26)	433 (49)
2E	210 (33)	436 (39)

(a) The first value reported is the peak center, and the value in brackets is the frequency at half-width maximum (FHWM).

The internal molecular vibrations for water and hydroxyl groups occur in the 150-500 meV energy range in the INS spectra, and these regions of the spectra for **1** and **2** are shown in Figures 1e and 1f. These spectra have been scaled such that the area under the O–H stretching regions (375–500 meV) are equal to two, corresponding to the two degrees of stretching vibration (symmetric and antisymmetric stretching modes). The peaks at ca. 200 meV in these spectra correspond to the H-O-H bending motion of molecular water. The position of this peak is consistent with the location of the H-O-H bending [$\delta(H_2O)$] mode observed in the spectrum of hydrated ZrO_2 nanoparticles.[23,24] This peak is not observed in the spectrum for **2B** due to the absence of molecular water in this dehydrated SnO_2 sample. The peaks associated with the internal vibrations of the molecular water groups have been modeled with Gaussian functions, the details of which are given in Table 2.[25] The peaks associated with the $\delta(H_2O)$ mode of molecular water occur at higher energies and with greater FHWM values in the spectra of the hydrated SnO_2 nanoparticles than the corresponding peaks in the spectra of the TiO_2 samples. This may be explained by disruption of the hydrogen bond network in the hydration layers on the TiO_2 nanoparticles relative to SnO_2, due to the stronger water-surface interactions in TiO_2. In addition, the larger FHWM values for the $\delta(H_2O)$ peaks in the TiO_2 spectra (**1**) with respect to the same peaks in the SnO_2 spectra (**2**) suggest greater dispersion of the H-O-H bending energy for water confined on the surface of TiO_2 relative to SnO_2. This is again consistent with stronger water-surface interactions in TiO_2 compared to SnO_2 that result in a greater distortion of the hydrogen bond network for the water.

The peaks at 410–436 meV in the spectra for the nanoparticle samples containing water (**1A**, **1B**, **2A**, **2C–2E**) can be assigned to the O-H stretching modes [$\upsilon_1(O–H)$] of molecular water, and are at energy values that are close to the related peak in the INS spectrum of hydrated ZrO_2

nanoparticles (436 meV).[24] The broadness of these peaks is again an indication of the large energy dispersion of this water vibrational mode due to the distortion of the hydrogen bond network in the water confined on the surface of these metal oxide nanoparticles.

A peak at 441 meV appears in the spectrum of **2B**. This peak corresponds to an O–H stretching vibration of the surface hydroxyl groups, and is presumably hidden in the other spectra by signals arising from the internal vibrations of the molecular water of the hydration layers. Alternatively, it is also possible that this hydroxyl stretching motion is less intense in these systems due to hydrogen bonding between the hydroxyl groups and the molecular water.

As shown in Table 2, the $\delta(H_2O)$ peaks appear at very similar energies in the spectra of **1A** and **1B**, and are located at similar energies in the spectra of **2C–E**. Equally, there is only a small difference between $\upsilon_1(O–H)$ peak energies for samples **2D** and **2E**. These observations imply that the internal vibrations of the molecular water confined to the surface of TiO_2 and SnO_2 nanoparticles are not significantly influenced by the level of hydration or size of the particles.

Additional peaks due to O–H stretching modes of the surface hydroxyl groups can be observed in all spectra between 500–520 meV. These peaks are broad and are due to two-phonon neutron scattering. The peak energies are not exact sums of the energies of the $\upsilon_1(O–H)$ and librational modes due anharmonicity of the systems. This peak appears at similar energies in the spectra of **1** and **2** because the variation in the positions of the librational bands in these spectra, due to differences in the hydrogen bond environments of the surface water, is compensated for by an opposing shift in the energy of the water stretching modes, and consequently the sum of these energies appears almost unchanged.

CONCLUSIONS

We have presented INS spectra for hydrated TiO_2 rutile and SnO_2 cassiterite nanoparticles from 0-550 meV, allowing a direct comparison of the dynamics and structure of water confined on the surface of these metal oxide particles. This study has shown that the translational motion of the confined water on these nanoparticles is greatly suppressed relative to ice-Ih and that this effect is more pronounced for H_2O on TiO_2 rutile compared to SnO_2 cassiterite. Further differences between the two metal oxide systems were detected in the librational regions of the INS spectra that indicate that the intermolecular hydrogen bond network within the hydration layers on TiO_2 rutile is more perturbed than the corresponding network in water confined to the surface of SnO_2 nanoparticles. The variations in the structure of the water confined on these isostructural metal oxide nanoparticles are consistent with the higher surface energy reported for nano-TiO_2 (2.22 ± 0.07 Jm^{-1}) compared to nano-SnO_2 (1.72 ± 0.01 Jm^{-1}).[12]

ACKNOWLEDGMENTS

N. L. Ross, E. C. Spencer, A. Navrotsky and A. A Levchenko acknowledge support from the U.S. Department of Energy, Office of Basic Energy Sciences (DOE–BES), grant DE FG03 01ER15237. A. I. Kolesnikov and D.L. Abernathy wish to acknowledge ORNL/SNS and W. Wang and D. J. Wesolowski acknowledge support from DOE-BES grant ERKCC41 (Nanoscale Complexity at the Mineral-Water Interface) at Oak Ridge National Laboratory that is managed by UT-Battelle, LLC, for DOE under contract DE-AC05-00OR22725. Research at the SNS at ORNL was sponsored by the Scientific User Facilities Division, Office of Basic Energy

Sciences, U.S. Dept. of Energy. Argonne National Laboratory is supported by DOE–BES under contract DE-AC02-06CH11357. We are thankful to L. Jirik for assistance with the INS experiments conducted at the IPNS. D. J. Wesolowski acknowledges J. Rosenqvisk and L. Anovitz at ORNL for their assistance with the cassiterite sample preparation and characterization.

REFERENCES

1. M.-I. Baraton, and L. Merhari, *J. Nanopar. Res.*, **6**, 107 (2004)
2. G. A. Waychunas, C. S. Kim, and J. F. Banfield, *J. Nanopar. Res.*, **7**, 407 (2005)
3. V. M. Aroutiounian, V. M. Arakelyan, and G. E. Shahnazaryan, *Solar Energy*, **78**, 581 (2005)
4. H. A. Al-Abadleh, V. H. Grassian, *Sur. Sci. Rep.*, **52**, 63 (2003)
5. G. Li, L. Li, J. Boerio-Goates, and B. F. Woodfield, *J. Am. Chem. Soc.*, **127**, 8659 (2005)
6. A. A. Levchenko, G. Li, J. Boerio-Goates, B. F. Woodfield, and A. Navrotsky, *Chem. Mater.* **18**, 6324 (2006)
7. J. Boerio-Goates, G. Li, L. Li, T. F. Walker, T. Parry, and B. F. Woodfield, *Nano Lett.*, **6**, 750 (2006)
8. C. Alba-Simionesco, B. Coasne, G. Dosseh, G. Dudziak, K. E. Gubbins, R. Radhakrishnan, and M. Sliwinska-Bartkowiak, *J. Phys.: Condens. Matter.*, **18**, R15 (2006)
9. M. Alcoutlabi, and G. B. Mckenna, *J. Phys.: Condens. Matter.*, **17**, R461 (2006)
10. The coherent (σ_{coh}), incoherent (σ_{inc}), and total (σ_{tot}) neutron scattering cross-sections for hydrogen are 1.76, 80.26, and 82.02 barns, respectively. For titanium σ_{coh} = 1.48 barns, σ_{inc} = 2.87 barns, and σ_{tot} = 4.35 barns. For tin σ_{coh} = 4.871 barns, σ_{inc} = 0.022 barns, and σ_{tot} = 4.893 barns. For oxygen σ_{coh} = 4.23 barns, σ_{inc} = 0.00 barns, and σ_{tot} = 4.23 barns. These values are taken from V. F. Sears, *Neutron News*, **3**, 26 (1992)
11. E. C. Spencer, A. A. Levchenko, N. L. Ross, A. I. Kolesnikov, J. Boerio-Goates, B. F. Woodfield, A. Navrotsky, and G. Li, *J. Phys. Chem. A*, **113**, 2796 (2009)
12. Y. Ma, R. H. R. Castro, W. Zhou, A. Navrotsky, *J. Mater. Res.*, *in press* (2011)
13. A. V. Bandura, J. D. Kubicki, and J. O. Sofo, *J. Phys. Chem. B*, **112**, 11616 (2008)
14. E. Mamontov, L. Vlcek, D. J. Wesolowski, P. T. Cummings, W. Wang, L. M. Anovitz, J. Rosenqvist, C. M. Brown, and V. Garcia Sakai, *J. Phys. Chem. C*, **111**, 4328 (2007)
15. S. Liu, Q. Liu, J. Boerio-Goates, B. F. Woodfield, *J. Adv. Mater.*, **39**, 18 (2007)
16. C.-K. Loong, S. Ikeda, and J. M. Carpenter, *Nuc. Instr. Meth. Phys. Res.*, **A260**, 381 (1987)
17. A. I. Kolesnikov, J.-M. Zanotti, and C.-K. Loong, *Neutron News*, **15(3)**, 19 (2004)
18. D. L. Abernathy, *Natiziario Neutroni E Luce Di Sincrotrone*, **13(1)**, 4 (2008)
19. J. Li and A. I. Kolesnikov, *J. Mol. Liq.*, **100**, 1 (2002)
20. J.-C. Li, J. D. Londono, D. K. Ross, J. L. Finney, J. Tomkinson, W. F. Sherman, *J. Chem. Phys.*, **94(10)**, 6770 (1991)
21. E. C. Spencer, N. L. Ross, S. F. Parker, A. I. Kolesnikov, B. F. Woodfield, K. Woodfield, M. Rytting, J. Boerio-Goates, A. Navrotksy, *J. Phys. Chem. A*, *submitted* (2011)
22. A. A. Levchenko, A. I. Kolesnikov, N. L. Ross, J. Boerio-Goates, B. F. Woodfield, G. Li, and A. Navrotsky, *J. Phys. Chem. A*, **111**, 12584 (2007)
23. C.-K. Loong, J. W. Richardson, Jr., and M. Ozawa, *J. Catalysis*, **157**, 636 (1995)
24. M. Ozawa, S. Suzuki, C.-K. Loong, and J. C. Nipko, *Appl. Sur. Sci.*, **121/122**, 133 (1997)
25. PeakFit v4.12, SeaSolve Software Inc., 1999–2003.

Mater. Res. Soc. Symp. Proc. Vol. 1352 © 2011 Materials Research Society
DOI: 10.1557/opl.2011.1133

X-ray Diffraction Investigations of TiO₂ Thin Films and Their Thermal Stability

Radomír Kužel[1], Lea Nichtová[1], Zdeněk Matěj[1], Zdeněk Hubička[2], Josef Buršík[3]
[1]Department of Condensed Matter Physics, Faculty of Mathematics and Physics, Charles University in Prague, Ke Karlovu 5, 121 16 Prague 2, Czech Republic
[2]Institute of Physics, Academy of Sciences of the Czech Republic, Na Slovance 2, 182 21 Prague, Czech Republic
[3]Institute of Inorganic Chemistry of the AS CR, v. v. i., 250 68 Rez 1001, Czech Republic

ABSTRACT

In-situ laboratory measurements in X-ray diffraction (XRD) high-temperature chamber and detailed XRD measurements at room temperature were used for the study of the thickness, temperature and time dependences of crystallization of amorphous TiO₂ thin films. The films deposited by magnetron sputtering, plasma jet sputtering and sol-gel method were analyzed. Tensile stresses were detected in the first two cases. They are generated during the crystallization and inhibit further crystallization that also depends on the film thickness. XRD indicated quite rapid growth of larger crystallites unlike the sol-gel films when the crystallites grow mainly by increasing of annealing temperature.

INTRODUCTION

Titania is well-known material of great interest because of its low cost, good chemical stability, nontoxicity, mechanical hardness and optical transmittance with high refractive index, and in particular because it is a unique material that connects two distinct photo-induced phenomena: photocatalytic activity and photo-induced superhydrophilicity after UV illumination [e.g 1-5]. However, desired properties are strongly influenced by phase composition (anatase, rutile, brookite), the crystallinity (amorphous, crystalline, nanocrystalline), and/or by their particular microstructure, presence of residual stresses in the films etc.

Usually, amorphous films do not have required properties and crystalline or nanocrystalline form is preferred. In principle, these can be prepared by finding of suitable deposition parameters or they can be obtained by annealing of amorphous films. At the same time, because of a need of deposition onto substrates like polymers, lower temperatures are required for the preparation. Therefore, temperature evolution of the film structure and microstructure is of high interest. There has been some attention given to the crystallization of amorphous titania powders and thin films [e.g. 6-8]. In our previous studies [e.g. 9-13], magnetron-deposited nanocrystalline and amorphous films of different thickness were investigated by X-ray diffraction at room temperature and by in-situ measurements in high-temperature chamber. A strong influence of film thickness on composition and crystallization of TiO₂ thin films was found. In very thin amorphous films the crystallization is significantly slower than in thicker layers.

In this contribution, these results are compared with similar studies of TiO₂ films prepared by other methods – plasma jet sputtering and sol-gel technique.

EXPERIMENT

Three kinds of films were studied. Most experiments were performed for the films deposited by dual magnetron equipped with two Ti(99.5) targets of 50 mm in diameter and supplied by a dc–pulsed Advanced Energy Pinnacle Plus+ 5kW power supply unit (PSU) operating in bipolar asymmetric mode [14] on glass and silicon (100) crystal plates (Sets A).

The second set (B) of films was prepared with the help of pulsed double hollow Ti cathodes plasma jet sputtering. The cathodes were placed in the ultra-high vacuum chamber (10−6 Pa) and the operating pressure level was set at 3.5 Pa. Argon with the flow rate of 80 sccm (standard cubic centimeter per minute) was used as a carrier gas passing through both jets with the same velocity. Oxygen (80 sccm) streams were supplied to the reactor via a lateral entry. Cylindrical nozzles (hollow cathodes) were made of pure Ti (internal diameter 5mm) and their distance from the surface was exactly 30 mm. The jets were connected to the supply of DC pulse current with frequency 2.5 kHz. Sputtered particles were blown with help of Ar out of the nozzle into the reactor and carried by the plasma jet towards the substrate. All layers were simultaneously deposited on single crystalline silicon (100). The titanium atoms reacted with oxygen only after hitting the film in this case.

The third set of films was prepared by the sol-gel method. Ti isopropoxide ($Ti(OCH(CH_3)_2)_4$, Sigma Aldrich), was used as metal precursor. Calculated amounts of $Ti(OCH(CH_3)_2)_4$ was dissolved in iso-butanol, mixed and heated for several hours. Subsequently a suitable amount of 2,2-diethanolamine (DEA) used as a modifier was added. All reactions and handling were carried out under dry nitrogen atmosphere to prevent reaction with air humidity. Polished Si(111) were used as substrates. Thin films were deposited by spin-coating technique (RC 8 Gyrset, KarlSuss France) and dried at 110°C in air for 60 minutes. This deposition cycle was repeated several times as long as films of desired thickness were obtained (Sets C).

X-ray diffraction measurements were performed on two diffractometers - Philips X'Pert MRD and Panalytical X'Pert Pro in parallel beam setup, 2θ scans with the angles of incidence in the range of 0.5-3°, parallel plate collimator in the diffracted beam and the Goebel mirror in the primary beam (Philips). Detailed measurement of residual stresses was carried out on the Eulerian cradle by the $sin^2\psi$ method for several peaks. In addition, X-ray reflectivity curves (film thickness, density, surface roughness) were measured. In-situ high-temperature measurements were performed in MRI high-temperature chamber with both radiant and direct heating mounted in vertical Panalytical X'Pert Pro theta-theta goniometer. Only a few anatase peaks were analyzed during the in-situ measurements. The peaks were fitted with the Pearson VII function and the peak positions, integrated intensities and integral breadths were determined. Some diffraction patterns taken in parallel beam geometry were evaluated by the total pattern fitting with the aid of the program MStruct developed in our laboratory (see [9]).

RESULTS AND DISCUSSION

Most of the results of studies on the sets A have already been published in [9-13]. The studies were performed on a set of as-deposited amorphous films with different thickness. By post-annealing, it was found that the amorphous TiO_2 films can be crystallized at about 250 °C [10]. This "temperature of fast crystallization" (the word "fast" means here the appearance of the diffraction peak within a few minutes) was confirmed also by isochronal annealing and it was determined to 220 °C for the films thicker than 300 nm. For thinner films this temperature increased up to 280 °C (for the thinnest film - 48 nm).

For isothermal studies, lower temperatures were selected than those of the so-called fast crystallization temperature [14]. Then, the crystallization still occurs but it is rather slow and can be studied in greater detail. The selected temperatures were mostly 180 °C. Large differences between crystallization rate and crystallization onset of thicker and thinner films can clearly be seen (Fig. 1). The experimental data could be well fitted the well-known Johnson-Mehl-Avrami-Kolmogorov (JMAK) equation [15, 16], modified as $I = 1-\exp[-k(t-t_0)^n)]$, where k is the rate of the process and n is the transformation index related to the dimensionality of the growth and on the kinetic order of nucleation, t_0 is the crystallization onset that had to be included in the equation for fitting. Experimentally, this corresponds to the first appearance of any significant diffracted intensity above background in the position of the expected diffracted peak. The integrated intensities I of the diffraction peaks are directly related to the crystallized fraction of the film with corresponding crystallographic orientation. Similar studies have already been performed with synchrotron radiation in [6].

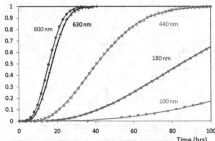 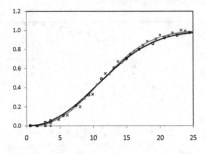

Figure 1. Dependence of normalized (to corresponding maximum values after full crystallization) integrated intensities of anatase XRD 101 peak on annealing time for films with different thickness at temperature of 180 °C (set A). The solid lines correspond to theoretical calculations with the JMAK equation.

Figure 2. Dependence of normalized integrated intensities of anatase XRD 101 (o, thin line), 004 (•, thick line) and 200 (x) peaks on annealing time at temperature 260 °C (set B). The solid lines correspond to theoretical calculations with the JMAK equation.

The set B differs from A mainly in lower film density as it can be detected by both X-ray reflectivity measurements (systematically smaller angle of total reflection by about 0.01 °) and significantly lower XRD intensities for the corresponding film thickness. Probably, this fact was reflected in higher crystallization temperature of these films. The isothermal experiments similar to set A could only be performed at temperatures higher than 250 °C. In Fig. 2, an example is shown for three anatase diffraction peaks. Some variations could be found but basic features of the dependence are the same for all the peaks (fitted lines correspond to $n = 2.39$ and 2.25 and $k = 0.0021$ and 0.0027 for 004 and 101 peaks, respectively, film thickness of about 600 nm.

The values of n slightly increase with the film thickness (set A) but it is in the range 2–2.5 for both sets and this is much lower value than that for 3D growth according the model. This indicates a lower dimension growth type. The crystallization onset t_0 increases rapidly with the decreasing thickness, while the rate k increases for the thicker films, see [14] for more. More detailed study of thickness dependence for set B is under way.

For both sets, it has been found that *XRD line broadening* is rather small, comparable with the instrumental broadening, from the very beginning of crystallization and does not vary with the annealing time. This indicates fast growth of relatively large crystallites (~ 100 nm and more) already at the beginning of crystallization as it has also been found in [6]. It must be noted that all the studies concern the directions somewhat inclined to the sample surface (~ 10-25° depending on the diffraction peak).

For both sets A and B, systematic variation of *XRD peak intensities* with the annealing time was also found. In set A, the most pronounced variations can be found for the ratio 101/004 that is increasing with the time up to certain saturated value corresponding roughly to random distribution of crystallites. This indicates that at the beginning of crystallization the grains oriented with the planes (001) parallel to the surface are preferentially developed but with the proceeding time this preferred orientation is suppressed (in the used 2θ scan, both planes studied are inclined to the surface by the angles of about 11° and 18°, respectively). However, the ratio for set B is completely different and it is decreasing with the time going even more apart from the random distribution.

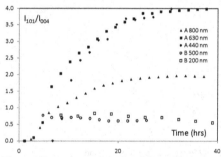 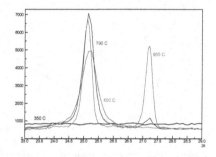

Figure 3. Dependence of I_{101}/I_{004} diffracted intensity ratio for anatase peaks on th annealing aling time for films with different thickness at temperature 180 °C (set A) and 260 C (set B).

Figure 4. Evolution of diffraction anatase peak 101, set C, with temperature during isochronal annealing (100 °C/h). The peak on the right corresponds to rutile.

Significant *shifts of diffraction peaks* towards higher values with the time until certain saturated value was reached (set A) were observed. This corresponds to the evolution of tensile stress during the annealing. Detailed studies of *tensile residual stresses* on post-annealed samples of set A measured at room temperature were performed by the analysis of the d-spacing variation with the lattice plane inclination for several diffraction peaks [12]. It was found that the stresses significantly decrease with increasing film thickness from about 300-400 MPa for the films thicker than 400 nm up to 1 GPa for 50 nm film. The results were confirmed by the total pattern fitting of 2θ scans [9]. For the films of set B, the situation was more complicated and no clear shifts were observed during the annealing. Finally, room temperature stress measurements indicated the presence of small tensile stresses (about 300 MPa) but the $\sin^2\psi$ dependences were not unique for all the peaks as in the case of set A. The differences are probably caused by smaller grain interaction in less dense films B.

Annealing of the sol-gel films (set *C*) showed behavior clearly different from the previous films. At lower annealing temperatures the diffraction peaks were very broad indicating small crystallite size and the broadening decreased with increasing temperature (Fig. 4). Phase transition into rutile appeared at about 790 °C unlike set *B* where no rutile was detected below 930 °C (max. used temperature). During isothermal annealing the diffracted intensity varied in a way similar to sets *A*, *B*, i.e. in agreement with the JMAK model (Figs. 5, 6) and *n* = 1.8. However, the peak broadening is decreasing with the time (Fig. 6). At each temperature a saturated value is reached after certain time. Then the temperature must be increased in order to further reduce the broadening. This behavior is similar to amorphous titania powders when predetermined crystallite size can be obtained by annealing of amorphous powders for specific time at specific temperature. The procedure cannot be well used for thin films of sets *A* and *B*. The XRD line broadening is assumed to be related more to crystallite size than microstrain. However, full pattern analysis is under way for the studied samples.

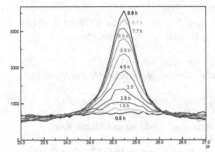

Figure 5. Evolution of diffraction anatase peak 101 with time (in hours) at 390 °C (set C).

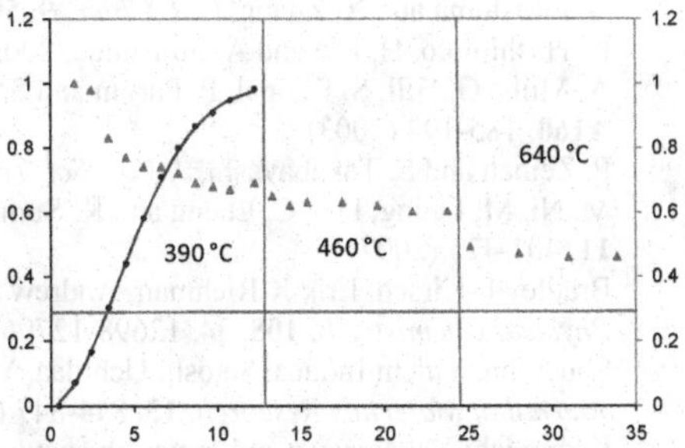

Figure 6. Dependence of normalized intensity of 101 anatase peak (dots, left axis, line corresponds to JMAK fit). Dependence of 101 peak FWHM (in degrees, triangles, right axis) on the annealing time (change of temperatures indicated). Horizontal line indicates the level of instrumental XRD line broadening.

CONCLUSIONS

XRD studies of crystallization of TiO_2 films showed that the process strongly depends on the deposition conditions and deposition method. Magnetron and plasma jet sputtered amorphous films crystallize at quite low temperatures. The process is determined not only by the temperature but the annealing time also plays an important role. During the crystallization, tensile stresses are generated that inhibit further crystallization and cause pronounced thickness dependence of the process. Higher stresses appear in more compact films. Evolution of preferred grain orientation during crystallization was observed as well. In magnetron deposited films the process ends up with nearly random crystallite distribution. The texture depends on the deposition method. As follows from narrow and constant XRD line broadening, relatively large crystallites (> 100 nm) are formed quickly from the very beginning of crystallization.

Annealing of amorphous sol-gel films leads to different results. Crystallization starts at higher temperatures by forming small crystallites which reach specific size at specific temperature after definite time. For further size reduction, the temperature must be increased.

ACKNOWLEDGMENTS

The work is supported by the Grant Agency of the Academy of Sciences of the Czech Republic under numbers KAN40072070, Grant Agency of the Czech Republic P108/11/1539 and also as a part of the research plan MSM 0021620834 financed by the Ministry of Education of the Czech Republic.

REFERENCES

1. A. Fujishima and X. Zhang, *C.R. Chim.* **9**, 50-760 (2006).
2. K. Hashimoto, H. Irie and A. Fujishima, 2005, *Jpn. J. Appl. Phys.*, **44**, 8269–8285 (2005)
3. A. Mills, G. Hill, S. Bhopal, P. Parkin and S. A. O'Neill, *J. Photochem. Photobiol.*, **A160**, 185-194 (2003).
4. P. Zeman and S. Takabayashi, *J. Vac. Sci. Technol.*, **A20**, 388-393 (2002).
5. M. Ni, M. Leung, D.Y.C. Leung and K. Sumathy, *Renewable Sustainable Energy Rev.*, **11**, 401-425 (2007).
6. Bradley L Kirsch, Erik K Richman, Andrew E Riley, Sarah H Tolbert, *Journal of Physical Chemistry B*, **108**, 34, 12698-12706 (2004).
7. Shu Yina, Yuichi Inouea, Satoshi Uchidaa, Yoshinobu Fujishiroa and Tsugio Satoa, *Journal of Materials Research*, **13**, 844-847 (1998).
8. Kazumichi Yanagisawa and James Ovenstone, *J. Phys. Chem. B*, **103** (37), 7781–7787 (1999).
9. Z. Matěj, R. Kužel and L. Nichtová, *Powder Diffraction*, **25**, 125-131 (2010).
10. R. Kužel, L. Nichtová, Z. Matěj, D. Heřman, J. Šícha and J. Musil, *Z. Kristallogr.* Suppl. **26**, 247-252 (2007).
11. R. Kužel, L. Nichtová, Z. Matěj and J. Musil, *Thin Solid Films,* **519(5)**, 1649-1654 (2010).
12. Z. Matěj, R. Kužel and L. Nichtová, *Metallurgical and Materials Transactions* **A**, (2011) in press, DOI: 10.1007/s11661-010-0468-z.
13. R. Kužel, L. Nichtová, Z. Matěj, J. Šícha and J. Musil, *Z. Kristallogr.*, **27**, 287-294 (2008).
14. P. Baroch, J. Musil, J. Vlcek, K. H. Nam and J. G. Han, *Surf. Coat. Technol.*, **193**, 107 (2005).
15. M. J. Avrami, *J. Chem. Phys.* **8**, 212 (1941).
16. A. N. Kolomogorov, *Izv. Akad. Nauk SSSR*, Ser. Mat. **3**, 355 (1937).

Mater. Res. Soc. Symp. Proc. Vol. 1352 © 2011 Materials Research Society
DOI: 10.1557/opl.2011.876

Morphology and Electronic Structures of Calcium Phosphate Coated Titanium Dioxide Nanotubes

Lijia Liu, Sun Kim, Jeffrey Chan and Tsun-Kong Sham
Department of Chemistry, University of Western Ontario
London, ON N6A5B7, Canada

ABSTRACT

Titanium dioxide nanotubes (TiO_2-NT) have been synthesized via an electrochemical anodization strategy followed by calcination under different temperatures to form TiO_2 nanostructures of anatase and rutile crystal phases. The nanotube-on-Ti structure is further used as a substrate for calcium hydroxyapatite (HAp) coating. The effect of TiO_2 morphology and crystal phases (i.e. amorphous, anatase and rutile) on the coating efficiency of HAp has been investigated in comparison with HAp coating on bare Ti metal. The HAp coated TiO_2-NT have been studied using X-ray absorption near-edge structure (XANES) at the Ti K- and Ca K-edge. The results show that TiO_2 of amorphous and anatase phases are of comparably good performance for HAp crystallization, and both are better than rutile TiO_2, while HAp is hardly found on bare Ti. The implications of the findings are discussed.

INTRODUCTION

Titanium has been widely recognized as an orthopedic implants material due to its light weight and high bio-compatibility [1]. However, due to its metallic property, the interaction between Ti and bones are not strong enough, which limits the practical application. Effort has been made to increase the surface area of Ti by decrease the size of Ti to small scales [2] or by growing a porous layer [3-4], which is typically oxide on Ti surface. The latter, which can be achieved by acid or alkali etching a Ti foil, provides a layer of porous TiO_2 or TiO_2 nanotubes directly attached to the metallic Ti substrate. It is known that TiO_2 has various crystal phases, among which anatase and rutile are most common. It is thus desirable to find out which crystal phase of TiO_2 could serve as a better interface for bone-implant interaction.

Calcium hydroxyapatite (HAp), $Ca_{10}(PO_4)_6(OH)_2$, is one of the calcium orthophosphates (Ca-P) compounds. It is of particular interest due to its high stability and low solubility, thus it becomes an ideal coating material of the metal implant in order to enhance the bone bonding and implant fixation [5]. HAp can crystallize directly from simulated body fluid (SBF) under appropriate conditions. However, due to the complicated crystallization process [6], whether or not the Ca-P compound crystallized from SBF is in fact, having the HAp structure, is still not entirely clear.

In this study, TiO_2-NT of amorphous, anatase, and rutile structure are synthesized via electrochemical anodization followed by calcination. Our previous study suggests that TiO_2-NT can crystallize from amorphous to anatase and further to rutile by controlling the calcination temperature [7]. The TiO_2-NT samples are thus used as substrates for HAp coating. The electronic structures of TiO_2-NT and HAp are examined using X-ray absorption near-edge structures (XANES) at Ti K- and Ca K-edge. XANES is an X-ray spectroscopic technique, which utilizes a tunable X-ray source (synchrotron radiation) to excite a core electron of an element of interest to its previously unoccupied electronic states. The absorption coefficient thus monitored is highly sensitive to chemical local environment. Experimentally, XANES is recorded in total electron yield (TEY) and X-ray fluorescence yield (FLY) simultaneously. It is a great tool for studying layered structures [8], since the two detection modes are of different sampling depth: TEY is surface sensitive and FLY is bulk sensitive.

EXPERIMENT

Electrochemical anodization was performed using a home-made electrochemical setup with a two-electrode configuration. Detailed description can be found elsewhere [7]. Ti foil (0.1 mm, Goodfellow Ltd.) was cut into 20 mm x 5 mm squares and used as anode and Pt wire was used as cathode. The electrolyte was 0.5 wt % HF and glycerol with a volumetric ratio of 1:9. Under a constant potential of 24 V for 6 hours at room temperature, Ti surface was self-organized into a nanotubular structure of TiO_2, henceforth denoted as TiO_2-NT-ap. TiO_2-NT-ap was then calcinated at 500°C and 800°C, respectively in order to form TiO_2 of the anatase and the rutile phase, respectively. The calcinated TiO_2-NT were henceforth denoted as TiO_2-NT-A for anatase and TiO_2-NT-R for rutile.

Ca-P was coated onto TiO_2-NT using a biomimetic deposition method [9]. TiO_2-NT-ap, TiO_2-NT-A, TiO_2-NT-R and bare Ti foil were used as substrates. Briefly, the substrate was dipped into a 0.02 M $Ca(OH)_2$ and a 0.02 M $(NH_4)_2HPO_4$ solution alternatively for 20 cycles. The immersion duration in each solution was 1 min, and the sample was rinsed with de-ionized water for 1 min every time prior to the next dipping step. The samples were then immersed into a simulated body fluid (SBF) for one day (24 hours) at 37°C. The SBF was prepared using standard method according to a reported procedure [10]. The Ca-P coated samples were denoted as CaP-NT-ap, CaP-NT-A, CaP-NT-R, and CaP-NT-Ti. The morphology of TiO_2-NT before and after Ca-P deposition was characterized using scanning electron microscopy (SEM, LEO 1540) at the Nanofarbrication lab of the University of Western Ontario.

Synchrotron measurements were conducted the at Soft X-ray Micro-characterization Beamline (SXRMB) of Canadian Light Source, University of Saskatchewan [11]. This beamline was equipped with a double crystal monochromator using either InSb (111) or Si(111) crystals. It was designed with the emphasis placed on the energy range of 2-5 keV although it can operate at higher energy. The samples were placed onto the sample holder using a double-sided carbon tape and the holder was then mounted on a 4-axis sample manipulator and was at a 45° angle with respect to the incident beam. XANES were recorded in the experimental chamber under high vacuum in total electron yield (TEY), which measures the specimen current, and total X-ray fluorescence yield (FLY), using a multichannel plate. All spectra were normalized to the incident photon flux.

DISCUSSION

Morphology

Figure 1 shows the top-view of TiO_2-NT before and after calcination. It can be seen that the as-prepared TiO_2-NT (Figure 1(a)) was vertically grown on the surface with diameters around 80 nm. Calcinated TiO_2-NT at 500°C and 800°C are shown in Figure 1(b) and 1(c), respectively. The morphology of TiO_2-NT after calcinations at 500°C remains the same as TiO_2-NT-ap, but after calcinations at 800°C, the nanotubes collapse and transform into micro-scaled fused columns. Their structures are confirmed with XRD in agreement with our previous studies, which have shown that TiO_2-NT undergoes phase transformation from amorphous to anatase and further to rutile under alleviated temperatures [7]. The electronic structures of these three TiO_2-NT samples related to their crystal phases will be presented in the following subsection.

Figure 1. SEM images of TiO_2-NT. a) TiO_2-NT-ap, b) TiO_2-NT-A, c) TiO_2-NT-R

Ca-P coating was conducted using the three above-mentioned TiO_2-NT and Ti foil as substrates. The SEM images of TiO_2-NT after immersion in SBF solution for 24 hours are shown in Figure 2. The CaP-NT-ap and CaP-NT-A are of very similar morphology, which form clusters with needle-shaped nanostructures pointing outwards. The Ca-P crystal growth starts from aggregating at certain sites on the surface and a secondary structure then crystallize around the initial nuclei. Between the CaP-NT-A and CaP-NT-ap, less aggregation is seen in CaP-NT-A with a more uniform coverage. The Ca-P deposited on TiO_2-NT-R substrate only forms individual particles on the surface, and the substrate is still clearly seen. It should be noted that no Ca-P is observed on bare metallic Ti substrate thus the SEM image is not shown.

Figure 2. SEM images of TiO_2-NT after Ca-P deposition. a) CaP-NT-ap, b) CaP-NT-A, c) CaP-NT-R

Electronic structures of TiO₂ and Ca-P from Ti and Ca K-edge XANES

We first take a look at the TiO$_2$-NT before and after calcination. Ti K-edge XANES of three TiO$_2$-NT samples recorded in TEY and FLY are shown in Figure 3(a). Due to the fact that the X-ray penetration depth at the Ti K-edge is much larger (i.e. 20 μm below the edge and 5 μm above the edge) [12] than the electron escape depth, the FLY signal is mainly from the metallic Ti substrate, especially in TiO$_2$-NT-ap and TiO$_2$-NT-A. TEY, however, clearly shows that after anodization, the surface of Ti turns into TiO$_2$, with absorption spectra significantly different from metallic Ti. Since K-edge XANES probes the unoccupied electronic states via Ti 1s → 4p dipole transition at the edge and the Ti 1s → 3d quadrupole transition (forbidden) in the pre-edge region, the Ti K-edge XANES of TiO$_2$ are strongly affected by crystal field thus can be used to distinguish between anatase and rutile crystal phases due to their differences in t$_{2g}$-e$_g$ splitting [13]. We can clearly see from Figure 3(a) that the spectra are composed of several weak peaks below the absorption threshold (pre-edge), followed by several resonance above the edge. The TiO$_2$-NT-ap only shows broad peaks at both regions, indicating its amorphous structure. After calcination, TiO$_2$-NT-A and TiO$_2$-NT-R exhibit well-defined peaks that correspond to the characteristic features of anatase and rutile, respectively [14-15]. At the pre-edge region, the spectrum has a mixed contributions from quadruple transition of 1s → 3d and dipole transition of 1s → p, in which TiO$_2$ of anatase phase has three peaks while the first peak weakens in rutile. Differences between anatase and rutile can also be seen at post-edge region, that is that rutile has two peaks between 4980 eV and 4995 eV while anatase only has one peak followed by some weak oscillations.

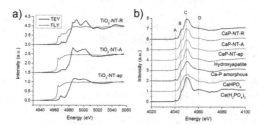

Figure 3. (a) Ti K-edge XANES of TiO$_2$-NT recorded in TEY and FLY. (b) Ca K-edge XANES of Ca-P on TiO$_2$-NT substrates samples in comparison with Ca-P standards. All spectra are normalized to unit edge jump.

The electronic structures of the deposited Ca-P were examined at the Ca K-edge by comparison with Ca-P standards. The spectra are shown in Figure 3(b). It can be seen that all Ca-P standards share several common features across the edge: 1) a weak pre-edge peak A at ~4040 eV, which corresponds to Ca 1s → 3d transition, 2) a shoulder B at ~4045 eV, which is attributed to the Ca 1s → 4s transition (via hybridization), 3) the main peak C at ~4050 eV, which is the 1s → 4p transition, and 4) several broad resonances at higher energy (D and higher) [16]. However, differences still exist in the spectra due to the slight variations of the Ca environment (e.g. Ca-O distances, coordination numbers, etc.), if we examine the peak profiles closely. Shoulder B is clearly seen in HAp, while broadened in amorphous Ca-P, and shifted toward higher energy side in $Ca(H_2PO_4)_2$; a shoulder next to peak C at the higher energy side of comparative intensity is seen in HAp and $Ca(H_2PO_4)_2$, indicating contribution from two types of Ca sitting at different sites of the lattice; but the amorphous Ca-P only shows one broad peak. In the case of $CaHPO_4$, complex features are seen that are different from other Ca-P samples: instead of a shoulder followed by a sharp peak, the main resonance is composed of several oscillations. Comparing the Ca-P on TiO_2 samples with the standards, it is immediately apparent that all Ca-P crystallized from SBF solutions have features identical to those of HAp regardless of substrates and crystal morphologies.

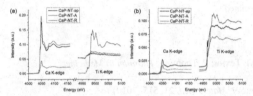

Figure 4. Ca K- and Ti K-edge XANES of CaP-NT-ap, CaP-NT-A and CaP-NT-R recorded in (a) TEY and (b) FLY.

The effect of substrate related to the thickness of the deposition is examined by comparing the XANES of Ca K- and Ti K-edges using TEY and FLY, which is of different sampling depth. TEY is dominant by signal from sample surface of a few nm, while FLY is more bulk sensitive. The edge jump (difference in absorption coefficient between pre-edge and post-edge) is proportional to the concentration of the element of interest in the material. In the case of CaP-NT-R, as observed from the SEM image, CaP only partially covers the surface of TiO_2, thus only TEY shows a small Ca K-edge jump relative to the other two samples and both TEY and FLY are dominated by Ti K-edge features of rutile. TEY of both CaP-NT-ap and CaP-NT-A exhibit a large edge jump at the Ca K-edge, but only weak and noisy response at Ti K-edge as expected, since TiO_2 is mostly covered by CaP as observed from SEM images. The FLY, on the other hand, shows signals from both surface CaP and TiO_2-NT underneath. Note that the sharp peak at 4966 eV is from the metallic Ti substrate. The edge jump ratio of Ca and Ti (I_{Ca}/I_{Ti}) is 0.217 for CaP-NT-ap and 0.151 for CaP-NT-A, indicating a thicker layer of CaP was formed on as-prepared TiO_2-NT than on anatase TiO_2-NT under same experimental conditions.

CONCLUSIONS

TiO$_2$-NT was synthesized using electrochemical anodization and calcination under different temperatures. This method produces TiO$_2$ of anatase and rutile crystal phases. The nanotubes were used as substrates and coated with calcium phosphate from a simulated body fluid. It is found that hydroxyapatite is successfully deposited from SBF onto all substrates including amorphous, anatase and rutile. Among the three, amorphous TiO$_2$ and anatase TiO$_2$ have significantly higher CaP coverage and better CaP crystalline texture than rutile TiO$_2$. Under the same experimental conditions, CaP layer on amorphous TiO$_2$ is slightly thicker than that on anatase TiO$_2$, but is more uniform on the latter.

ACKNOWLEDGMENTS

Research at UWO is supported by NSERC, CFI, OIT and CRC. The Canadian Light Source is supported by NSERC, NRC, CIHR and the University of Saskatchewan. Technical support from Dr. Yongfeng Hu, beamline scientist of SXRMB beamline is greatly appreciated.

REFERENCES

1. D. M. Brunette; P. Tengvall; M. Textor; P. Thomsen, *Titanium in Medicine*; Berlin: Springer, (2001).
2. T. Reiner; I. Gotman, *J. Mater. Sci. Mater. Med.* **21**, 515 (2010).
3. K. Das; S. Bose; A. Bandyopadhyay, *J. Biomed. Mater. Res. A* **90A**, 225 (2009).
4. H.-M. Kim; T. Kokubo; S. Fujibayashi; S. Nishiguchi; T. Nakamura, *J Biomed. Mater. Res.* **52**, 553 (2000).
5. S. V. Dorozhkin; M. Epple, *Angew. Chem. Int. Ed.* **41**, 3130 (2002).
6. L. Wang; G. H. Nancollas, *Chem. Rev.* **108**, 4628 (2008).
7. L. Liu; J. Chan; T.-K. Sham, *J. Phy.Chem. C* **114**, 21353 (2010).
8. X. T. Zhou; T. K. Sham; W. J. Zhang; C. Y. Chan; I. Bello; S. T. Lee; H. Hofsass, *J. Appl. Phys.* **101**, 013710/1 (2007).
9. A. Kodama, A.; Bauer, S.; Komatsu, A.; Asoh, H.; Ono, S.; Schmuki, P. *Acta Biomaterialia* **5**, 2322 (2009).
10. T. Kokubo; H. Kushitani; S. Sakka; T. Kitsugi; T. Yamamuro, *J. Biomed. Mater. Res.* **24**, 721 (1990).
11. Y. F. Hu; I. Coulthard; D. Chevrier; G. Wright; R. Igarashi; A. Sitnikov; B. W. Yates; E. L. Hallin; T. K. Sham; R. Reininger, *AIP Conf. Proc.* **1234**, 343 (2010).
12. X-ray calculator, http://henke.lbl.gov/optical_constants/atten2.html
13. M. F. Ruiz-Lopez; A. Munoz-Paez, *J. Phys. Condens. Matter* **3**, 8981 (1991).
14. R. Brydson; H. Sauer; W. Engel; J. M. Thomas; E. Zeitler; N. Kosugi; H. Kuroda, *J. Phys. Condens. Matter* **1**, 797 (1989).
15. Z. Y. Wu; J. Zhang; K. Ibrahim; D. C. Xian; G. Li; Y. Tao; T. D. Hu; S. Bellucci; A. Marcelli; Q. H. Zhang; L. Gao; Z. Z. Chen, *Appl. Phys. Lett.* **80**, 2973 (2002).
16. D. Eichert; M. Salome; M. Banu; J. Susini; C. Rey, *Spectrochimica Acta, B* **60B**, 850 (2005).

Mater. Res. Soc. Symp. Proc. Vol. 1352 © 2011 Materials Research Society
DOI: 10.1557/opl.2011.758

Comparison between the optical and surface properties of TiO₂ and Ag/TiO₂ thin films prepared by sol-gel process

Marcelo M. Viana and Nelcy D. S. Mohallem*

Laboratory of Nanostructured Materials, Chemistry Department, UFMG, Belo Horizonte-MG, Brazil
*nelcy@ufmg.br

ABSTRACT

Colloidal precursor solutions, obtained from a mixture of titanium isopropoxide, isopropyl alcohol and silver nitrate, were used to fabricate amorphous TiO₂ and Ag/TiO₂ thin films by sol-gel process. The films were deposited on borosilicate substrates, which were heated at 400 °C for 30 minutes and cooled rapidly to the formation of amorphous coatings. The films were investigated by X-ray diffraction, scanning electron microscopy, atomic force microscopy and UV-vis spectroscopy. The thickness, roughness, refraction index, and particle size of the TiO₂ and Ag/TiO₂ films were determined and compared. Finally, hydrophobic-hydrophilic property was evaluated to the thin films produced.

INTRODUCTION

Systems containing TiO₂ nanoparticles have attracted interest in several research areas by providing numerous applications in different industries, linked to improving the quality of life. As example we can mention applications in pharmaceuticals, pigments, food and cosmetics, as well as others related to the semiconducting properties of TiO₂, with emphasis on the photocatalytic applications, mainly involving the photodegradation of organic pollutants [1-4].

Systems containing Ag nanoparticles also have been investigated mainly due to their potential applications in modern electronic devices [5], medicine [6], optical sensors [7] and biology [8].

The preparation of composites containing both the materials favors the coupling of their structures, leading to additional or enhanced properties. These phenomena can be verified in their photocatalytic properties [9] and in the surface plasmon resonance (SPR) formed by the Ag nanoparticles distribution on the TiO₂ surface [10] that can modify the optical, hydrophobic-hydrophilic and bactericidal properties of the material surface.

In this work, optical, structural and morphological properties of TiO₂ and Ag/TiO₂ thin films prepared by sol-gel process were evaluated using X-ray diffraction (XRD), scanning electron microscope (SEM), atomic force microscopy (AFM) and UV-visible spectroscopy. Moreover, TiO₂ and Ag/TiO₂ thin films were tested for their hydrophobic- hydrophilic properties.

EXPERIMENT

TiO$_2$ thin films have been prepared by sol-gel process from an alcoholic solution containing tetraisopropyl orthotitanate (Merck) and isopropyl alcohol (Merck) and the Ag/TiO$_2$ films by using the same precursors and silver nitrate (Aldrich). The atomic ratio Ag:Ti was 1:6. UV- C irradiation (λ=254 nm) were used to promote the Ag$^+$ reduction in solution. The precursor solutions were maintained under agitation at room temperature for 1 hour and the films were dip-coated onto clean glass substrates with withdrawal speed of 8 mm/sec, dried in air for 30 min and thermally treated for 15 min at 400 °C followed by fast cooling.

The structure of the TiO$_2$ and Ag/TiO$_2$ thin films were observed by low angle X-ray diffractometry (incidence angle of 5°) at the D12A-XRD1 beam of the LNLS/Campinas using radiation of 1.5424Å. The transmittance of the films was determined by optical transmission spectra obtained with an ultraviolet and visible spectrometer (U3010 – Hitachi). The morphology and texture of the films were analyzed by atomic force microscopy (Asylum-MFP-3D) and scanning electron microscopy (SEM, type Quanta 200 FEG - FEI) using an accelerating voltage of 30 kV.

Hydrophobic/hydrophilic tests were carried out, dropping water onto thin films to measure the contact angle in the surfaces, and irradiating the film surfaces with UV light.

DISCUSSION

The TiO$_2$ and Ag/TiO$_2$ thin films produced show amorphous character since no diffraction peak concerning to TiO$_2$ and metallic Ag have been observed in the diffractograms (Figure 1). This result evidences that the rapid cooling used produced thin films of TiO$_2$ with atomic-scale disorder. The Ag nanoparticles on the TiO$_2$ surface could not be detected due to their nanometric size and distribution on the surface.

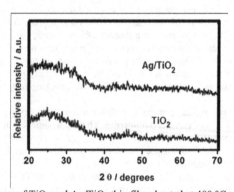

Figure 1. X-ray patterns of TiO$_2$ and Ag/TiO$_2$ thin films heated at 400 °C and rapidly cooled.

Figure 2 shows SEM images of topography and cross-section of the thin films produced. The TiO$_2$ film has a flat and homogeneous surface with low porosity while the Ag/TiO$_2$ film shows silver nanoparticles with approximately spherical morphology dispersed on its surface.

The silver nanoparticles segregate from the TiO$_2$ matrix throughout its porous network during heating and migrate mainly to the surface of the thin film [9]. The size distribution of silver nanoparticles ranges from 2 to 40 nm and presents two size domains centered at 7 and 25 nm as shown in inset of Figure 2a. The estimated thicknesses to TiO$_2$ and Ag/TiO$_2$ films were of 150±15 and 250±25 nm, respectively.

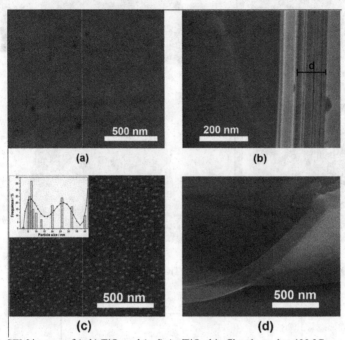

Figure 2. SEM images of (a-b) TiO$_2$ and (c-d) Ag/TiO$_2$ thin films heated at 400 °C.

Figures 3a and b show that the TiO$_2$ films are formed by nanoparticles with sizes of approximately 10 nm, confirming the homogeneity of the film surface in according to SEM results. The surface roughness (RMS) of the TiO$_2$ film is 0.55 nm, and the insertion of silver leads to an increase in its surface roughness (RMS=2.30 nm). The silver nanoparticles observed in Figure 3c and d present sizes in accordance to particle size distribution shown in Figure 2c.

Transmission spectra of the thin films are shown in Figure 4. The transmittance of the TiO$_2$ thin film is higher than that to Ag/TiO$_2$ in the entire range of wavelength. This result is related to light absorption by silver nanoparticles. The refractive index of 2.03 (λ=550 nm) to Ag/TiO$_2$ thin film is lower than that obtained to TiO$_2$ thin film (n=2.33). Besides, by comparison of the transmission curves, it could be seen that the absorption edge shifts to higher wavelengths for the Ag/TiO$_2$ suggesting a reduction in the TiO$_2$-band gap value. This observation is related to formation of an effective silver/titania interface that benefits the photogenerated electron transfer between Ag and TiO$_2$.

Figure 3. AFM images of (a-b) TiO$_2$ and (c-d) Ag/TiO$_2$ thin films heated at 400 °C.

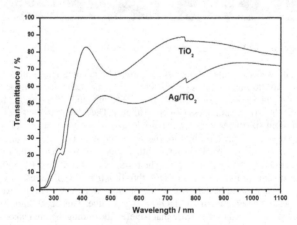

Figure 4. Optical transmission spectra of TiO$_2$ and Ag/TiO$_2$ thin films.

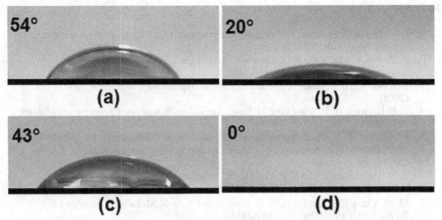

Figure 5. Photographs of water droplet shape on the amorphous (a) TiO_2 and (b) TiO_2 after UV-C irradiation, (c) Ag/TiO_2 and (d) Ag/TiO_2 after UV-C irradiation.

The UV-light-induced hydrophilicity for both the films is shown in Figure 5. When a water drop contacts the TiO_2 and Ag/TiO_2 thin film surfaces, the contact angles observed are 54° (Figure 5a) and 43° (Figure 5c), respectively. Upon UV light irradiation of the film surfaces, a water drop dripped on these surfaces spread out resulting in contact angles of 20° and 0° to TiO_2 (Figure 5 b) and Ag/TiO_2 (Figure 5d) thin films, respectively. These results demonstrate that the hydrophobic character of the films changed to superhydrophilic one in the prepared films, and that this effect is more accentuated when silver nanoparticles are present. When the films return to the dark, they recover their hydrophobic behavior, showing that the photo-induced hydrophilic process is reversible.

CONCLUSIONS

Amorphous TiO_2 and Ag/TiO_2 thin films heated at 400 °C were produced successfully by sol-gel methodology. Ag/TiO_2 films showed higher thickness and roughness values since as the silver nanoparticles migrate to surface of the thin film. The surface wettability demonstrated a reversible hydrophobicity-hydrophilicity transition when UV light was used to irradiate the film surface. The silver insertion in the TiO_2 films modified the optical and surface properties of the obtained films.

ACKNOWLEDGMENTS

This work was supported by Brazilian funding agencies CNPq, FAPEMIG and Petrobrás. We thank the Microscopy Centre of UFMG by SEM and AFM images and the LNLS/Campinas by the use of the XRD1 beam.

REFERENCES

1. K. Eufinger, D. Poelman, H. Poelman, R. Gryse and G. B. Marin, *Journal of Physics D: Applied Physics* **40**, 5232 (2007).
2. P. Bouras, E. Stathatos, P. Lianos and C. Tsakiroglou, *Applied Catalysis B: Environmental* **51**, 275 (2004).
3. L. Zhao, Y. Yu, L. Song, X. Hu and A. Larbot, *Applied Surface Science* **239**, 285 (2005).
4. T. L. Hsiunga, H. P, Wanga and H. Ping, *Journal of Physics and Chemistry of Solids* **69**, 383 (2008).
5. F. Xue, Z. Liu, Y. Su, K. Varahramyan, *Microelectronic Engineering* **83**, 298 (2006).
6. A. R. Shahverdi, A. Fakhimi, H. R. Shahverdi and S. Minaian, *Nanomedicine* **3**, 168 (2007).
7. A. D. Mcfarlandand and R. P. Van Duyne, *NanoLetters*, **3** (8), 1057-1062 (2003).
8. H. Huangand and X. Yang, *Carbohydrate Research* **339**, 2627 (2004).
9. M. M. Viana, C. C. De Paula, D. R. Miquita and N. D. S. Mohallem, *J. Sol-Gel Sci. Technol.* DOI 10.1007/s10971-011-2455-2
10. U. Kreibig and M. Vollmer, *Optical Properties of Metal Clusters*, (Springer, Berlin, 1995).

Mater. Res. Soc. Symp. Proc. Vol. 1352 © 2011 Materials Research Society
DOI: 10.1557/opl.2011.1080

Optical and electrical properties of solution processable TiOx thin films for solar cell and sensor applications

Jiguang Li[1], Lin Pu[2] and Mool C. Gupta[1*]
[1] Department of Electrical and Computer Engineering, University of Virginia, Charlottesville, Virginia 22904, U.S A.
[2] Department of Chemistry, University of Virginia, Charlottesville, Virginia, 22904, U.S.A.

ABSTRACT

Recently, few tens of nanometer thin films of TiO_x have been intensively studied in applications for organic solar cells as optical spacers, environmental protection and hole blocking. In this paper we provide initial measurements of optical and electrical properties of TiO_x thin films and it's applications in solar cell and sensor devices. The TiO_x material was made through hydrolysis of the precursor synthesized from titanium isopropoxide, 2-methoxyethanol, and ethanolamine. The TiO_x thin films of thickness between 20 nm to 120 nm were obtained by spin coating process. The refractive index of TiO_x thin films were measured using an ellipsometric technique and an optical reflection method. At room temperature, the refractive index of TiO_x thin film was found to be 1.77 at a wavelength of 600 nm. The variation of refractive index under various thermal annealing conditions was also studied. The increase in refractive index with high temperature thermal annealing process was observed, allowing the opportunity to obtain refractive index values between 1.77 and 2.57 at a wavelength 600 nm. The refractive index variation is due to the TiO_x phase and density changes under thermal annealing.

The electrical resistance was measured by depositing a thin film of TiO_x between ITO and Al electrode. The electrical resistivity of TiO_x thin film was found to be 1.7×10^7 Ω.cm as measured by vertical transmission line method. We have also studied the variation of electrical resistivity with temperature. The temperature coefficient of electrical resistance for 60 nm TiO_x thin film was demonstrated as $- 6 \times 10^{-3}/°C$. A linear temperature dependence of resistivity between the temperature values of $20 - 100$ °C was observed.

The TiO_x thin films have been demonstrated as a low cost solution processable antireflection layer for Si solar cells. The results indicate that the TiO_x layer can reduce the surface reflection of the silicon as low as commonly used vacuum deposited Si_3N_4 thin films.

INTRODUCTION

Titanium dioxide (TiO_2) thin films have been extensively used in optical applications, such as antireflection coatings, waveguides and so on [1-3]. Various methods can be used for the fabrication of the TiO_2 thin films such as PECVD [4], sputtering [5] etc. Sol-gel method is one of the most promising technique because it is a low cost and simple process [6, 7].

Heeger's group reported that a novel so-gel TiO_x thin film could increase the efficiency and life time of organic solar cells. The synthesis and properties of TiO_x material is different from typical TiO_2 prepared by sol–gel processes. The TiO_x thin films composition contains Ti:O = 42.1 : 56.4 as determined by X-ray photoelectron spectroscopy (XPS) measurement. The measured electron mobility was found to be higher than amorphous oxide films prepared by typical sol–gel processes [8, 9]. The high electron mobility of TiO_x thin film is due to the interpenetrating -Ti-O-Ti-O- network. The TiO$_x$ sol-gel precursor was synthesized without any acid catalyst which is necessary in typical TiO$_2$ sol-gel synthesis [10-14]. The photochemically activated TiO$_x$ protection mechanism for organic solar cells has been demonstrated in our previous study [15], indicating a good oxygen barrier properties.

The spectral threshold energy for TiO$_x$ thin films is 3.9 eV as we previously reported [15]. It is higher than typical sol-gel produced TiO$_2$ material [10, 14]. The applications of TiO$_x$ thin films in inorganic solar cells as well as other optical and photonic devices have not been explored. In this paper, we present the results on the variation of refractive index of TiO$_x$ thin films measured through ellipsometry and modeling. It is of interest to investigate the electrical properties of TiO$_x$ thin films. In this paper we present the initial results on the electrical resistivity of TiO$_x$ thin films. Vertical transmission line measurement method was used to extract the TiO$_x$ electrical resistivity.

The unique optical and electrical properties of TiO$_x$ synthesized without acid catalyst needs to be further investigated and compared with typical TiO$_x$ films made by sol-gel product.

EXPERIMENT

The TiO$_x$ material is synthesized according to reference [8, 9]. Process starts with the injection of titanium isopropoxide (Aldrich, 99.999%, 2 ml), 2-methoxyethanol (Aldrich, 99.9+%, 8 ml) and ethanolamine (Aldrich, 99.5+%, 0.8 ml) one by one into a three-necked flask. The mixed solution was stirred and heated at room-temperature, 80 °C and 120 °C for 1 hour each. The flask was supplied with continuous dry N$_2$ during all 3 hours. Then 4 ml methanol was injected to extract the final TiO$_x$ sol–gel precursor.

For optical study, the TiO$_x$ precursor solution was then spin-coated on the P-type silicon substrate. The sample was baked at 90 °C for 10 minutes so that the precursor was converted to TiO$_x$ by hydrolysis in air. Annealing was carried out in rapid thermal annealing system under O$_2$ for 10 minutes at different temperature.

For electrical resistivity study, TiO$_x$ sol gel was spin coated on indium tin oxide coated glass (Delta technology) substrate. Various spinning speed were used to achieve different film thicknesses. Then 80 nm aluminum thin film was deposited by e-beam evaporation on top of the TiO$_x$ after hydrolysis process.

The reflectance of the TiO$_x$ film on Si was measured by Filmetrics Filmeasure tool. The measurement of refractive index was carried out using Jobin Yvon UVISEL Ellipsometer. Optical constant modeling was performed using Jobin Yvon DeltaPsi2 software.

The electrical resistivity properties were calculated from I-V characteristic curve using a Keithley 2611 source meter. Temperature dependence of electrical resistivity measurements were made in a chamber filled with N$_2$. The chamber was designed with thermal control and temperature monitor.

RESULTS AND DISCUSSION

Figure 1 show variation of the refractive index changes at a wavelength of 600 nm when the sample is annealed in O$_2$. The data show that the refractive index increases with the annealing temperature. The refractive index changes are due to the removal of the organic residue material and densification of the film.

A double layer model was used to determine the refractive index of the film by ellipsometric method. The model consisted of a dense bottom TiO_x layer and a 50% TiO_x : 50% void surface layer. The TiO_x/void layer was used to simulate surface roughness. The dispersion model to describe the TiO_x material is based on Forouhi-Bloomer formula [16].

The refractive index of TiO_x was found to be 2.08 at 600 nm after 400 °C annealing in O_2. It is close to the commonly used vacuum deposited Si_3N_4 film, which has refractive index of 2.01 at 600 nm wavelength. The reflectance values at 600 nm wavelength for Si substrate, Si_3N_4 and TiO_x coated on Si substrate at 600 nm wavelength are shown in Table I. The table indicates that solution processable TiO_x material could be used as effective anti-reflection layer for silicon based solar cells.

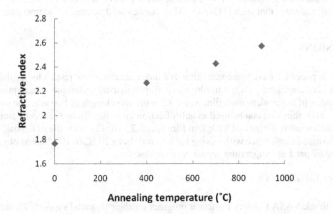

Figure 1. The refractive index at 600 nm wavelength for TiO_x annealed at various temperatures in O_2 environment

Table. I. Reflectance for Si substrate coated with various thin films at 600 nm wavelength

	Si substrate only	Si_3N_4 (film thickness=70 nm)	TiO_x thin film after 400 °C annealing in O_2 (film thickness=60 nm)
Reflectance	35%	2%	6 %

Shockley [17] proposed the original transmission line measurement method for calculation of electrical resistivity for thin films. We used thin films of TiO_x of various thicknesses and extracted the contact and bulk electrical resistance.

The extracted electrical resistivity of as deposited TiO_x thin film was 1.7×10^7 Ω.cm. For a device with 80 nm TiO_x, the total series specific resistance was 140 Ω.cm^2 with TiO_x bulk specific resistance as 136 Ω.cm^2, which is the major component of total series resistance.

We studied the electrical resistivity of ITO-TiO_x-Al device structure from 20 °C to 100 °C. The resitivity of TiO_x film decreased with raising temperature due to generation of additional carriers by thermal process. Table II shows the electrical resistance measurement results for 60 nm TiO_x thin film.

Table II. Variation of electrical resistance with temperature

Temperature (°C)	20 °C	40 °C	60 °C	80 °C	100 °C
Resistance (Ω)	1540	1380	1240	1070	863

The temperature coefficient of electrical resistance was calculated as -6.25×10^{-3} /°C. The good linearity indicates that such ITO-TiO_x-Al structure could be used for temperature sensor system.

CONCLUSIONS

In this paper we have presented refractive index measurement results for a solution processable semiconducting TiO_x thin film using ellipsometrical technique. The obtained refractive index of the as-deposited film was 1.82 at the wavelength of 600 nm. The solution processable TiO_x thin film can be used as antireflection layer for silicon solar cells similar to Si_3N_4. The electrical resistivity of TiO_x thin film was 1.7×10^7 Ω.cm for the as deposited sample. A linear decrease in resistance with raising temperature from 20 °C to 100 °C was observed, which could be used in temperature sensor applications.

ACKNOWLEDGMENTS

We thank NASA Langley Professor program for their financial support. We thank Professor John Yates for helpful discussions.

REFERENCES

1. A. Szeghalmi; M. Helgert, R. Brunner, F. Heyroth, U. Gösele, M. Knez, *Appl. Opt.* **48**, 1727-1732 (2009).
2. N. Nobuhiro, S. Mitsunori, M. Mitsunobu, B. Nobuyoshi and S. Naruhito, *Appl. Opt.* **30**, 1074-1079 (1991).
3. B. O'Reagan and M. Gräzel, *Nature.* **353**, 737-740 (1991).
4. P. R. McCurdy, L. J. Sturgess, S. Kohli, E. R. Fisher, *Appl. Surf. Sci.* **233**, 69-79 (2004).
5. S. Chao, W. Wang, M. Hsu, L. Wang, *J. Opt. Soc. Am. A* **16**, 1477-1483 (1999).
6. S. Phadke, J. D. Sorge, H. Sherwood, D. P. Birnie, III, *Thin Solid Films* **518**, 5467–5470 (2010).
7. M. M. Rahman, G. Yu, K. M. Krishna, T. Soga, J. Watanabe, T. Jimbo, and M. Umeno, *Appl. Opt.* 37, 691–697 (1998).
8. J. Y. Kim, S. H. Kim, H. H. Lee, K. Lee, W. Ma, X. Gong, A. J. Heeger, *Adv. Mater.* **18**, 572-576 (2006).
9. K. Lee, J. Y. Kim, S. H. Park, S. H. Kim, S. Cho, A. J. Heeger, *Adv. Mater.* 19, 2445-2449(2007).
10. B. R. Sankapal, M. Ch. Lux-Steiner, A. Ennaoui, *Appl. Surf. Sci.* 239, 165–170 (2005).

11. W. Kaewwiset, W. Onreabroy and P. Limsuwan, *Nat. Sci.* 42, 340 - 345 (2008).
12. M. Zaharescu, M. Crisan and I. Musevic, *J. Sol-Gel Sci. Technol.* 13, 769–773 (1998).
13. N. Ozer, H. Demiryont and J. H. Simmons, *Appl. Opt.* 30, 3661-3666 (1991).
14. P. Chrysicopoulou, D. Davazoglou, Chr. Trapalis, G. Kordas, *Thin Solid Films.* 323, 188–193 (1998).
15. J. Li, S. Kim, S. Edington, J. Nedy, S. Cho, K. Lee, A. J. Heeger, M. C. Gupta and J. T. Yates, Jr., *Sol. Energy Mater. Sol. Cells.* 95, 1123–1130 (2011).
16. A. R. Forouhi, I. Bloomer, *Phys. Rev. B.* 34, 7018-7026 (1986).
17. W. Shockley, *Report No. A1-TOR-64-207*, Air Force Atomic Laboratory, Wright-Patterson Air Force Base, Ohio, September 1964.

Mater. Res. Soc. Symp. Proc. Vol. 1352 © 2011 Materials Research Society
DOI: 10.1557/opl.2011.1342

Ellipsometric characterization of thin nanocomposite films with tunable refractive index for biochemical sensors

P. Petrik[1], H. Egger[2], S. Eiden[2], E. Agocs[1,3], M. Fried[1], B. Pecz[1], K. Kolari[4], T. Aalto[4], R. Horvath[1], D. Giannone[5]

[1]MFA, 1121 Budapest, Konkoly Thege u. 29-33, Hungary.
[2]Bayer Technology Services GmbH, 51368 Leverkusen, Germany.
[3]University of Pannonia, 8200 Veszprem, Egyetem u. 10, Hungary.
[4]VTT, P.O. Box 1000, FI-02044, Finland.
[5]Multitel, 2, rue Pierre et Marie Curie, B-7000 Mons (Belgium).

ABSTRACT

Creating optical quality thin films with a high refractive index is increasingly important for waveguide sensor applications. In this study, we present optical models to measure the layer thickness, vertical and lateral homogeneity, the refractive index and the extinction coefficients of the polymer films with nanocrystal inclusions using spectroscopic ellipsometry. The optical properties can be determined in a broad wavelength range from 190 to 1700 nm. The sensitivity of spectroscopic ellipsometry allows a detailed characterization of the nanostructure of the layer, i.e. the surface roughness down to the nm scale, the interface properties, the optical density profile within the layer, and any other optical parameters that can be modeled in a proper and consistent way. In case of larger than about 50 nm particles even the particle size can be determined from the onset of depolarization due to light scattering. Besides the refractive index, the extinction coefficient, being a critical parameter for waveguiding layers, was also determined in a broad wavelength range. Using the above information from the ellipsometric models the preparation conditions can be identified. A range of samples were investigated including doctor bladed films using TiO_2 nanoparticles.

INTRODUCTION

In order to fabricate highly sensitive biosensor chips based on polymer photonic crystal micro-cavities, two polymeric materials with a refractive index difference as large as possible are necessary. Even though polymers themselves usually do not have a very high or low refractive index, their values can be manipulated by preparing nanocomposite materials. Incorporation of high index materials into polymers is intensively studied and it has been shown that the refractive index of the original polymer material can be increased significantly. One way is to incorporate nanoparticles with a very high refractive index (like specific inorganic oxides) without losing the other desired properties of the polymers. This is also the most promising way when considering the processability of the composite materials. Therefore, the optimal particles concerning material, size, etc. is identified to maximize the refractive index while obtaining a convenient nanocomposite material.

The present work is a part of the European P3SENS project [1] that aims at the design, fabrication and validation of a multichannel polymer photonic crystal-based label-free disposable biosensor. The photonic chip will be based on polymer photonic crystal micro-cavities coupled on a planar waveguide optical distribution circuit. The photonic chip will be fabricated by emloying cost-effective fabrication technologies with an emphasis on low cost substrates (polymers) and replication technologies (nano-imprint lithography). It has been demonstrated that grating coupled waveguide biosensors [2-4] and photonic crystal devices [5-8] can be fabricated using polymers and replication technologies. However, the relatively low refractive index of the polymers limits their applicability as light guiding layers in biosensor applications. To overcome this limitation and achieving higher index contrast the lowering of the refractive index of the waveguide substrates were proposed using freestanding polymer films [2] or nanoporous supports with refractive index around 1.2 [3,4]. When polymer photonic crystal are fabricated the lack of full photonic band gap when immersed into water is again a serious limitation. This is due the low refractive index contrast (the refractive index of the polymer is in the range of 1.6 and that of the water is ~1.33). However, high index polymers (n ≈ 1.8) using embedded nanoparticles could allow the fabrication of photonic crystal polymer waveguides with a photonic band gap even when immersed in water, opening up several innovative directions.

In this article we present the development of optical models for the ellipsometric characterization of polymer thin films with refractive indices tuned by nanoparticle inclusion. First, demonstrate the sensitivity of spectroscopic ellipsometry [9,10] for such nanoparticles thin film structures. After, we present the measured refractive indices and extinction coefficients as a function of the weight fraction of nanoparticles.

EXPERIMENT

Polyimide (PI) has been chosen as the waveguide core material due to its low extinction coefficient (10^{-4}) and high intrinsic refractive index (1.64). TiO_2 nanoparticles (with refractive indices around 2.5) were used to increase the refractive index of PI. The nanocomposite dispersions were prepared according to the following procedure. Dispersions of TiO_2 nanoparticles in isopropanol with an approximately mean particle size of 50 nm in length have been transferred into N- ethylpyrrolidone (NEP) by dialysis. This final dispersion had a solid content of 11.2 wt%. This dispersion has been mixed with the PI solution, also in NEP, in different ratios. . After ultrasonic treatment stable dispersions are obtained, which showed a good compatibility of the TiO_2 nanoparticles with the PI solution. The TiO_2 dispersion was added to the PI solution to obtain PI-to-TiO_2 ratios of 0, 9:1, 4:1, 2:1, 1:1 and 1:1.5 resulting in dry films with TiO_2 weight percents (wt%) of 0, 10, 20, 33, 50 and 60. Moreover 10 wt% of butoxyethanol has been added to improve the wetting of the dispersion on the silicon substrate. The final dispersions had a PI content of 3.125 wt%.

The PI films with the TiO_2 nanoparticles were created using doctor blading with a spiral blade and a slit size of 10 μm onto blank silicon wafers. Using this technique we could obtain layer thicknesses in the range of 300-400 nm with a good homogeneity. Transmission electron microscopy images reveal that the nanoparticles (with sizes of approximately 10 nm) are homogeneously distributed within the layer (Fig. 1).

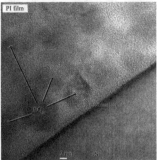

Figure 1. Transmission electron microscopy images showing the uniform distribution of the nanoparticles (left-hand side) and the TiO_2 nanoparticles in high resolution (right-hand side).

The ellipsometric measurements were performed using a Woollam M2000DI in situ rotating compensator ellipsometer that is capable of recording full spectra in the wavelength range from 191 to 1690 nm within about 1 s. This speed and the capability of measuring in a relatively small spot (about 0.3 mm by 0.6 mm) make this tool perfectly suitable with a high sensitivity detailed mapping of even larger surface areas. In this study we mapped a surface of 2 cm by 2 cm with a typical resolution of 5 points by 5 points.

In the ellipsometric evaluations we used two optical models. The simple one (Model 1 in Fig. 2) applies a single top layer for the PI film. The refractive index of this layer is calculated using the Cauchy parameterization of $n = A + B/\lambda^2 + C/\lambda^4$ and $k = De^{-E\lambda/\lambda 0}$, where n and k are the refractive index and the extinction coefficient, respectively. λ denotes the wavelength, whereas A, B, C, D, and E are the Cauchy parameters. In Model 2, there is a layer below the Cauchy layer using an effective medium mixture of the Cauchy material and void (i.e. the wavelength-independent refractive index of 1). The Cauchy components in the two layers are coupled, i.e. we use the same parameters for their refractive index. The refractive index of the c-Si reference was taken from the Woollam database [11].

Structure	Model 1	Model 2
PI film	Cauchy	Cauchy
		Cauchy + voids
Si	c-Si	c-Si

Figure 2. Optical models used for the ellipsometric characterizations.

DISCUSSION

Model 2 (see Table I for the fitted parameters on Sample 129) allows an excellent fit (Fig. 3) for all of the measured samples using 8 fit parameters. The reliability of these optical models is proven by the fact that in spite of the relatively large number of fit parameters the uncertainties (confidence limits shown by the ± values) are small for all the parameters (orders of magnitude smaller than the fitted values). This shows that our method fulfilled the requirements of increasing the number of fit parameters: a further parameter may be introduced if (i) it is physically relevant, (ii) it reduces the fit error (the mean squared error, MSE, typically reduces from 20 to 10 when switching from Model 1 to Model 2), and finally (iii) the parameter correlations are small (shown by the small uncertainties).

Table I. Parameters of Model 2 (see Fig. 1) fitted onto the measurement of the sample with a TiO₂ wt% of 33. f_v denotes the volume fraction of 'voids'. Layer 1 is the top (surface) layer.

Layer	Parameters
1	Thickness: 334.5.6±1.5 nm A = 1.704±0.001, B = 0.0142±0.0003, C = 0.00058±0.00004 D = 0.0079±0.0004, E = 1.17±0.11
2	Thickness = 11.3±1.3 nm f_v = 55.2±3.9 %

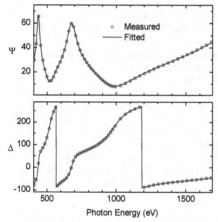

Figure 3. Typical fit of the calculated ellipsometric spectra onto the measured ones for the sample with TiO₂ wt% of 33. Ellipsometry measures the ratio of the reflection coefficients parallel and perpendicular to the plane of incidence, called the complex reflection coefficient. Ψ denotes the amplitude and Δ denotes the phase of this complex number. The mean squared error is 13. A value around 10 denotes already a very good fit.

The fact that the fit quality can significantly be improved using a two-layer model with a sublayer of modified optical density shows that there might be some vertical distribution of the nanocrystals. However, the thickness of this layer is in all cases in the range of 10-20 nm or even below, showing a high optical quality film. Note, that the void component in Layer 2 (see Table I) does not necessarily mean 'voids' in the layer. Rather, it shows the above mentioned (optical) density deficit at the interface as compared to the 'bulk' layer.

The high quality surface is revealed by the fact that no surface roughness layers were needed to obtain an excellent fit quality. When the surface roughness was fitted, we obtained values changing from 1 nm to 6 nm with increasing weight percent of TiO_2 from 0 % to 60 %, in good agreement with atomic force microscopy (AFM) results (although the AFM vs. SE comparison is not straightforward [12]).

The measuring speed of the Woollam M2000DI ellipsometer allows to make lateral maps with a good resolution within a reasonable time. The results revealed a good lateral homogeneity in terms of both layer thickness and refractive index (Fig. 4).

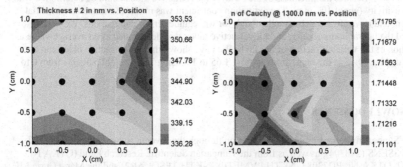

Figure 4. Lateral maps of the thickness of Layer 1 (of Table I) and the refractive index (n, calculated at the photon wavelength of 1300 nm) on the sample with a TiO_2 wt% of 33.

The refractive index can be modulated in the range from 1.65 to approximately 1.8 when varying the weight percent of TiO_2 nanoparticles in the PI solution. As expected, the refractive index varies in a broader range for smaller wavelengths (Fig. 5), however, the extinction coefficient is smaller for larger wavelengths. The most important wavelength is 1300 nm for waveguide application (this case is shown on the right-hand side of Fig. 5).

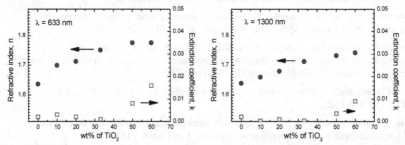

Figure 5. The mean value of the refractive index and extinction coefficient range of the lateral maps as a function of the weight fraction of TiO_2 in the PI solutions for the photon wavelengths of 633 nm and 1300 nm. The error of the measurements is smaller than the symbol size in all cases (see Table I for typical uncertainties).

CONCLUSIONS

We have created optical models for the ellipsometric characterization of polymer layers with tuned refractive indices using TiO_2 nanoparticles. We have shown that the optical model can significantly be improved by taking into account a lower index boundary layer adjacent to the substrate. The reason of this effect might be a slightly non-uniform vertical distribution of the nanoparticles. Besides the thickness of the PI layer the refractive index could be determined with high sensitivity (about 10^{-4} in n). The extinction coefficient was modeled using an exponential function, and it was found to be small enough for waveguide sensor applications. We have also checked the lateral homogeneity, and the refractive index was determined as an average value as a function of the volume fraction of the TiO_2. We have shown that the refractive index can be modulated by ≈ 0.15 in the range from approx. 1.65 to 1.8 using TiO_2 weight percents from 0 to 60 %.

ACKNOWLEDGMENTS

Support from the European Commission through the seventh framework program FP7-ICT4-P3SENS (248304), as well as from the Hungarian Scientific Research Fund (OTKA Nr. K81842, OTKA Nr. PD73084, NKTH PVMET08, NKTH TFSOLAR2) and the Marie Curie EIF Reintegration Fellowship (Opticell) is gratefully acknowledged.

REFERENCES

[1] http://www.p3sens-project.eu/

[2] Horvath R, Lindvold LR, Larsen NB, "Fabrication of all-polymer freestanding waveguides" Journal of Micromechanics and Microengineering 13(3), 419-424 (2003).

[3] Horvath R, Pedersen HC, Larsen NB, "Demonstration of reverse symmetry waveguide sensing in aqueous solutions" Applied Physics Letters 81(12), 2166-2168 (2002).

[4] Horvath R, Pedersen HC, Skivesen N, Svanberg C, Larsen NB, "Fabrication of reverse symmetry polymer waveguide sensor chips on nanoporous substrates using dip-floating" Journal of Micromechanics and Microengineering 15 1260-1264 (2005).

[5] Brian Cunningham, "Cunning Novel biosensors from photonic crystals" SPIE 2008 10.1117/2.1200801.1022.

[6] Choi, C.J. and Cunningham, B.T., "Single-step fabrication and characterization of photonic crystal biosensors with polymer microfluidic channels" Lab Chip 6, p1373 (2006).

[7] Ciminelli, C. et al, "Modelling and design of a 2D photonic crystal microcavity on polymer material for sensing applications", Third European Workshop on Optical Fibre Sensors, Proceedings of SPIE Vol. 6619, 661933.

[8] Kee, C-S and al. "Photonic band gaps and defect modes of polymer photonic crystal slabs", Appl. Phys. Lett. 86 (2005) 051101.

[9] Harland G. Tompkins, Eugene A. Irene (eds), Handbook of Ellipsometry; William Andrew, Springer, 2005.

[10] M. Fried, T. Lohner, P. Petrik, in: Ellipsometric Characterization of Thin Films, in: H.S. Nalwa (Ed.), Handbook of Surfaces and Interfaces of Materials: Solid Thin Films and Layers, Academic Press, San Diego, CA, 2001, p. 335 (Chapter 6).

[11] C. M. Herzinger, B. Johs, W. A. McGahan, and J. A. Woollam, "Ellipsometric determination of optical constants for silicon and thermally grown silicon dioxide via a multi-sample, multi-wavelength, multi-angle investigation", J. Appl. Phys. 83(1998) 3323.

[12] Petrik P, Biro LP, Fried M, Lohner T, Berger R, Schneider C, Gyulai J, Ryssel H "Comparative study of surface roughness measured on polysilicon using spectroscopic ellipsometry and atomic force microscopy", Thin Solid Films 315 (1998) 186.

Mater. Res. Soc. Symp. Proc. Vol. 1352 © 2011 Materials Research Society
DOI: 10.1557/opl.2011.790

Influence of Post-annealing Temperature on the Properties of Ti-Doped In₂O₃ Transparent Conductive Films by DC Ratio-frequency Sputtering

Lei Li, Chen Chen, Chengjun Dong, JiaJia Cao, Jingmin Dang, Yiding Wang*

State Key Laboratory on Integrated Optoelectronics, College of Electronic Science and Engineering, Jilin University, Changchun 130012, Jilin, P.R. China

ABSTRACT

In this paper, titanium doped (2 wt. %) indium oxide (TIO) thin films deposited on quartz substrates by DC sputtering were presented. Dealt with different temperatures from 420℃ to 620℃ of post-annealing in vacuum for 40 minuets, the samples display different optical and electric properties. The deposited films exhibited polycrystalline in the preferred (222) and (440) orientation, with higher mobility (up to 48.6 cm^2/VS) and lower resistivity ($1.26 \times 10^{-4}\Omega\cdot cm$) at the post-annealing temperature of 520℃. The average optical transmittance of the films is over 92% in a wavelength range from 300 to 1100 nm and the transmittance has only around 1.8% change with different post-annealing temperatures.

INTRODUCTION

Transparent conductive oxide (TCO) films have been attracted more and more attentions due to their high transmittance, low resistivity, high mobility and low cost. Their superior optical and electronical properties make them to be an excellent candidate for heterojunction solar cells, flat panel display devices, organic light emitting diodes and thin film transistors. These applications covered two tendencies in the research of TCO films. One is to improve the optical transparency in the near infrared (NIR) region. It can effectively extend the energy conversion efficiency in solar cells and coefficient of function in NIR lasers and detectors. And the other is to develop lower resistivity and higher transparency in the visible region. It can enlarge the display size in LED screens [1]. According to Drude model, without compromising the conductivity, the transparency window will shift to NIR region with decreasing carrier density as well as increasing the mobility of carriers [2].

Within the TCO films family, Al-doped zinc oxide, Tin-doped indium oxide and indium-doped cadmium oxide films have been reported frequently but the reports of Ti-doped indium oxide are seldom. Compared with other TCOs, the Ti-doped indium oxide films' optoelectronic properties are much higher, such as the high mobility and transmittance. Hest et al. have employed combinational deposition and analysis techniques to study Ti-doped In₂O₃ as a function of Ti doping concentration, substrate temperature and O₂ pressure (ρ: 6260Ωcm; μ: 80 cm^2/VS; n: $8.0 \times 10^{20}cm^{-3}$). Recently, Gupta et al. prepared Ti-doped indium oxide by pulsed laser deposition (PLD) and observed a mobility of 159-199 cm^2/VS and a minimum resistivity of $9.8 \times 10^{-5}\Omega cm$ [3-5].

The Ti-doped In_2O_3 reported in above literatures seems to have been adjusted by growth temperature, oxygen pressure and Ti concentrations. Generally, in order to obtain more excellent characteristics of polycrystalline films heat treatment is carried out. In this paper we are reporting on the influence of post-annealing temperature on the property of thin films which are prepared under the atmosphere of fixed O_2/Ar ratio of 1/12 and substrate temperature of 328℃.

EXPERIMENT

The films of titanium-doped In_2O_3 were deposited on quartz substrates by DPS-III ultra-high vacuum against-target RF sputtering. A sintered ceramic high purity In_2O_3 (99.99%) target embedded with 2 wt. % metal titanium, commercially available, was employed. The target dimensions were 60mm in diameter and 5mm in thickness. The quartz substrates were cleaned with de-ionized water in an ultrasound bath for 20 minuses before they were put into the chamber. The deposition chamber was evacuated to 1.0×10^{-4} Pa and the substrates temperature was kept at 328℃, while the mixture ratios of argon (99.999%) and oxygen(99.999%) were fixed at 24sccm and 2sccm. Prior to deposition, the sputtering current was launched at 150mA and the target was pre-sputtered in O_2+Ar ambient condition for 3-5mins to remove the impurity on the surface of the target. After 40 min sputtering, the samples were annealed in a quartz tube furnace at three different temperatures as 420, 520 and 620℃ for 30 min, and then the furnace was cooled to room temperature(RT).

The component and crystallinity were analyzed by a Bruker axs D8 advance X-ray power diffractmeter (XRD) with Cu-Kα radiation (λ = 1.5418Å) and an energy dispersion X-ray spectroscopy (EDX). The optical transmittance was measured by UV-1700, Shimadu spectrophotometer. Electrical resistivity, carrier concentrations and Hall coefficient measurement were carried out by Bio-Rad Micro science HL5500 Hall System. The surface morphologic was obtained by a field emission scanning electron microscopy (FESEM).

DISCUSSIONS

The optical transmittance of TIO films deposited on quartz at different post-annealing temperatures in vacuum are depicted in Fig. (1). It can be obverted that there is a little increase for the transmittance by alternating the post-annealing temperatures from room temperature to 620℃. The average optical transmittance of all films is over 92% and the highest transmittance(~96% at 540nm) shown for the film was dealt at 520℃. Furthermore, the absorption edge is shifted towards longer wavelength with the growing of annealing temperature because the carrier concentrations decrease which can be calculated according to the Drude model [2]:

$$\lambda_p = \frac{2\pi}{c}(\frac{m^* \varepsilon_0}{ne^2})^{1/2} \tag{1}$$

where λ_p is plasma wavelength, ε_0 is the free space permittivity, n is the carrier concentration, e is the charge, and m^* is the effective free electron mass. The transmittance windows, however, shift to longer wavelength can also be explained by the optical band-gap. When the carrier concentration decreases the band gap also decreases, which lead to the absorption edge increase. I will discuss this phenomenon in detail in the following part.

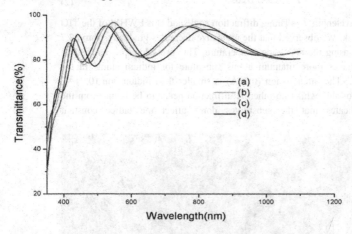

Figure 1. Optical transmission spectra for the TIO films with different post-annealing temperature: (a) as-deposited, (b) 420℃,(c) 520℃,(d) 620℃.

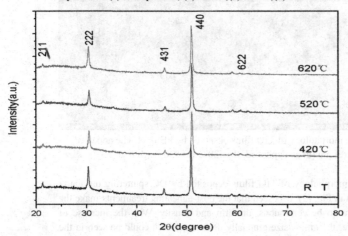

Figure 2. X-ray diffraction patterns of TIO films prepared at various post-annealing temperatures.

The XRD patterns of TIO films as-deposited and post annealing at different temperatures of 420, 520 and 620℃ are depicted in Fig. (2). It is obvious that the TIO films are polycrystalline, which the XRD spectra peak locations match those of pure indium oxide perfectly (MDI Jade 5.0). These films' preferential orientation is (440) and relative peaks are (211), (222), (431) and (622). The crystalline size from full-width at half-maximum (FWHM) value of (440) peak can be calculated by the Scherrer equation [6]:

$$D = 0.9\lambda / B\cos\theta \qquad (2)$$

where λ is X-ray wavelength, θ is Bragg diffraction angle and B is FWHM of the TIO (440) diffraction peak. We observed that the grain size decreases gradually from 22.7 to 20.5 nm by increasing the annealing temperature. This procedure proves that the high temperature makes more titanium atoms substitute for indium atoms in the indium oxide lattice. The titanium ion (0.74Å) is smaller than indium ion (0.94Å), which makes this possible whilst no other Ti diffraction peaks to be found from the XRD patterns indicates that the substitutes don't affect the lattice constant significantly.

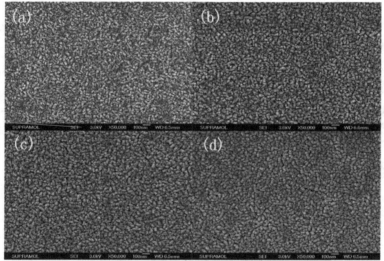

Figure 3. Surface morphology of TIO films observed by FESEM. The annealed temperatures are (a) as-deposited, (b) 420℃,(c) 520℃,(d) 620℃.

The surface morphology of TIO films deposited by DC sputtering on quartz substrates are shown in Fig. (3). It is seen that the post-annealing treatments make the surface of TIO films to be continuous, uniform and density. With the increase of annealed temperature, the grains size gradually decrease, which could be seen in the images of SEM but not very obvious. This decrease can attribute to the Ti ions observing enough diffusion energy to substitute for In ions in the crystal lattice and grains at higher temperature. The result of XRD proves this phenomenon.

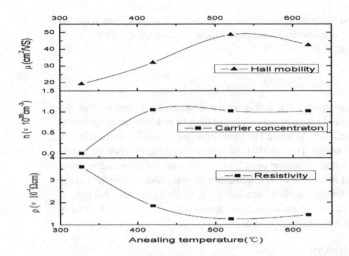

Figure 4. The resistivity, carrier concentration and Hall mobility as a function of different annealing temperature.

The changes in resistivity (ρ), carrier concentration (n) and Hall mobility (μ) as a function of annealing temperature are depicted in Fig. (4). The lowest resistivity of $1.26\times10^{-4}\Omega$cm was obtained at annealing temperature of 520℃. With the temperature increasing, the resistivity firstly decreases and reaches the lowest value and then increases at higher temperature of 620℃. While the Hall mobility show the absolutely different trend as the variation in resistivity and it reaches to the peak value of 48.6 cm²/VS at 520℃ from initial 19.2 cm²/VS(as-deposited).The carrier concentration increases with the annealing temperature and reaches to a constant value 1.02×10^{20} cm^{-3}.

It is generally believed that the negative sign of Hall coefficient proves the films are n-type semiconductors whose carriers are provided by O vacancies and doped ions with different valences. All films were prepared in ambient of O_2/Ar before the treatments of post annealing in the vacuum, and this makes the films exist more O vacancies while one vacancy can provide two extra electrons which will exist in the state of weakly bonding. Because of the doped ions and substituted ions with different valences, extra electrons also will be appeared when the doped ions substitute for substituted ions existing in the matrix oxides. These extra electrons will escape from the doped ions to be free carriers at high temperature [2, 7-8].

The changes of resistivity and carriers concentration with the increase of annealed temperature may be explained as follows: after Ti^{4+} substitute for In^{3+} in In_2O_3, one extra electron scatter to the gap between grains or occupy interstitial positions, at the same time the oxygen vacancies also can provide two extra electrons. All of these make the films' resistivity gets the minimum value. With the increase of temperature, these weakly bonded electrons observe energy to escape from the ions result in increasing carriers' concentration. These explanations are consistent with the result reported by R.K. Gupta et al.

CONCLUSIONS

This paper has reported the influence of post-annealing temperature on the properties of Ti-doped In_2O_3 transparent conductive films by ratio-frequency sputtering. The result showed that 2 wt% Ti doped TIO films deposited on quartz substrates at 328 ℃ and at the O_2/Ar ambient. Through the treatment of annealing, the films were well crystallized with preferred orientation of (222) and (440) and the grain size of about 21nm at orientation of (440). The average transmittance of the films is over 92% in the visible region. The lowest resistivity reaches to $1.26\times10^{-4}\Omega$cm with the highest mobility 48.6 cm^2/VS at the post-annealing temperature of 520 ℃. The carrier concentration of 1.02×10^{20} cm-3 was obtained.

ACKNOWLEDGEMENTS

Research supported by the National High Technology Research and Development Programs of China ("863"Programs), No.2007AA03Z112, No.2007AA06Z112, No. 2009AA03Z442, the Program of Jilin Provincial Science and Technology Department of China, No. 20090422 and the National Science Foundation of China, No.61077074.

REFERENCES

[1] T.Koida, M. kondo, J.Appl.Phys.101 (2007)063713.

[2] M.Yang, J.Feng, G. Li, Q. Zhang, J. Crystal Growth 310(2008) 3474

[3] R.K. Gupta, K. Ghosh, S.R. Mishra, P.K. Kahol, Appl. Surf. Sci. 253 (2007) 9422.

[4] R.K. Gupta, K. Ghosh, S.R. Mishra, P.K. Kahol, Mater.Lett.62 (2008) 1033.

[5] H.F.A.M. Hest, M.S. Dabney, J.D. Perkins, D.S. Ginley, M.P. Taylor, Appl. Phys. Lett.87 (2005) 032111.

[6] Z.B. Fang, Z.J. Yan, Y.S. Tan, X.Q. Liu, Y.Y. Wang, Appl. Surf. Sci. 241 (2005) 303.

[7] Z. Zhao, D. L. Morel, C. S. Ferekides, The Solid Films 413(2002) 203.

[8] G. Frank, H. Kostlin, Appl. Phys. Lett. A: Solids Surf. 27(1982) 197.

Applications

Mater. Res. Soc. Symp. Proc. Vol. 1352 © 2011 Materials Research Society
DOI: 10.1557/opl.2011.1009

Photocatalytic activity, hydrophilic and optical properties of nanocrystalline titania thin films prepared by sol-gel dip coating

M.C. Ferrara[1], L. Pilloni[2], A. Mevoli[1], S. Mazzarelli[1], L. Tapfer[1]

[1] ENEA, Brindisi Materials Technology Technical Unit (UTTMATB), Brindisi Research Centre, Strada Statale 7 Appia, 72100 Brindisi, Italy

[2] ENEA, Materials Technology Technical Unit (UTTMAT-CHI), Casaccia Research Centre, Via Anguillarese 301, 00060 S. M. di Galeria, Rome, Italy

ABSTRACT

Nanocrystalline anatase titania thin films were prepared by using two different precursor solutions, a highly acid solution (Sol-1) and a polymer-like solution (Sol-2), via the dip-coating technique on different substrates (<100>-Si wafer, fused silica and soda lime glass). The influence of the two sol-gel titania precursor solutions and of the substrate type on the film morphology, coating porosity, surface roughness, crystalline phases and grain size of the titania films were investigated. Our experimental results clearly indicate that the sol - composition and substrate type remarkably influence the microstructural/morphological properties of the titanium dioxide. They consequently modify the optical response and hydrophilic performances of the samples. The photocatalytic oxidations of the methylene blue in water of the samples grown on the glass substrate were monitored to investigate the influence of the sol-gel precursor solution on the photocatalytic activity of the titania coatings, and the results were put in relation with the hydrophilic and optical properties of the films. The outcome demonstrates that the optical properties and the hydrophilic and photocatalytic performances of nanocrystalline titania can be opportunely tailored tuning the size dimension of the crystalline domain according to the specific coating applications.

INTRODUCTION

Nanocrystalline TiO_2 anatase coatings are attracting much interest for their various industrial application such as photocatalytic degradation of air and water pollutants [1, 2] and for self-cleaning or antifogging materials [3, 4]. We have recently reported on the synthesis of nanocrystalline titania thin films by using two different sol-gel routes, i.e. different acid catalyzed sol-gel titania precursor solutions, via the dip-coating technique on (100)-Si wafer, fused silica and soda lime glass substrates [5]. In particular, the influence of two sol-gel titania precursor solutions, a TiO_2 nanoparticles bath, Sol-1, and a polymer titania sol, Sol-2, on the morphological, microstructural, optical and hydrophilic properties of TiO_2 films were investigated in detail. The results demonstrated that the sol composition and the substrate type have an influence on the structural, morphological and hydrophilic properties of the coating properties. We found that the coatings grown by using the polymeric sol are made of grains constituted of crystalline aggregated forms whereas the ones generated from the nanoparticulate bath were single crystals. The experimental results have also shown that the films prepared from the polymeric sol have the lowest mass density (about 70% of the theoretical bulk value of anatase), and exhibit super-hydrophilic properties, whereas the layers grown by the nanoparticulate sol have a much higher mass density (about 90-96% of the theoretical value), refractive index and contact angles. These results suggest that the sol composition plays a crucial role in the determination of the microstructural/morphological properties of the titanium dioxide, and that super-hydrophilic titania coatings can be obtained without UV irradiation or doping. This finding is a very important and interesting result for catalytic applications where a high capability of holding adsorbed water is preferable. On the basis of the results reported in literature we hypothesized that the high hydrophilic properties of our samples could be ascribed to the small size of the crystalline domains and the presence of oxygen vacancy on the surface of the titania grains that gives rise to Ti^{3+} sites and, consequently, to structural changes/defects of the anatase nano-architecture. The small crystalline size and the presence of defects on the titania crystallites were also considered to explain the optical results. In particular, we ascribed the high values of the energy gap (between 3.78 and 4.13 eV) of the samples to the quantum size effect due the small TiO_2 crystallite size (less than 14 nm in diameter). Furthermore, the observed band structure mutations from indirect to direct can be attributed to the presence of oxygen vacancy on the surface of the titania grains that gives rise to Ti^{3+} sites and, consequently, to the structural changes/defects of the anatase nano-architecture.

The energy gap values (E_g), as determined by optical reflectivity measurements, the mean crystalline domain size ($<D_C>$), the surface roughness (σ_{rsm}) and the mass density (ρ_m), as evaluated from GXRD, XSR and XDS measurements, the mean particle diameter (G) and the mean porosity diameter $<D_P>$, as determined by evaluating the digitalized FE-SEM images, and the contact angle ϑ of the TiO_2 films are reported in Table I for the samples prepared by nanoparticulate bath (Sol-1) and polymer sol (Sol-2) deposited on different substrates (Si(100) wafer, silica and glass).

Table I - Physical parameters of the TiO_2 films prepared by using nanoparticulate bath (Sol-1) and polymer sol (Sol-2) and deposited on different substrates.

precursor solutions	substrate	E_g (eV)	$<D_C>$ (nm)	$<D_P>$ (nm)	σ_{rsm} (nm)	ρ_m (g/cm³)	G (nm)	contact angle ϑ (°)	
								film	sub.
Sol-1	Si(100)	-	14±1	2±1	0.95±0.03	3.83	11±4	30±1	56
	silica	3.78± 0.01	12±1	2±1	1.00±0.05	3.56	11±4	29±1	5
	glass	3.90± 0.01	11±1	4±1	2.50±0.05	3.17	11±5	55±1	11
Sol-2	Si(100)	-	3.3	7±3	1.25±0.02	3.00	16±8	6±1	56
	silica	3.93± 0.01	4 6	7±3	1.30±0.03	2.85	16±8	7±1	5
	glass	4.13± 0.01	-	17±6	1.40±0.03	3.10	22±10	5±1	11

Successively, and this is the topic of this brief report, we investigated the photocatalytic activity of the coatings generated by using the two different titania sol-gel precursor solutions. In particular, we examined the effect of the different morphological/microstructures and contact angles of the coatings grown on soda lime glass substrate on their photocatalytic performance by investigating the photocatalytic oxidation of a solution of methylene blue (MB) in water due to the photocatalytic activity of the titania coatings illuminated by a UV-Vis lamp [6].

EXPERIMENT

Sol-gel synthesis and characterization techniques

Two titania precursors sols, a TiO_2-nanoparticles bath (Sol-1) and a polymer-like precursors solution (Sol-2) were prepared using titanium-isopropoxide (TIP) as alkoxide precursor of titanium (Ti), isopropanol (IPA) as solvent, H_2O for the hydrolysis and HCl as acid catalyst of the hydrolysis/condensation reaction of the alkoxide groups. Acetylacetone was employed in the polymeric-like titania solution as chelating agent of the titanium-alkoxide. The films were deposited under controlled temperature and atmospheric conditions (about 20°C and 31% relative humidity) by dip-coating techniques on (100)-Si wafer, fused silica and soda lime glass. The resulting gel coatings were first dried at 110°C for 20 minutes and then densified by means of a thermal treatment in air at 500°C for 1.5 hours.

The surface morphology, the grains size and the porosity of the samples were analyzed by a field emission scanning electron microscopy (FE-SEM LEO 1530). The microstructure (crystalline domain size), surface roughness and electron density of the titania films were measured by X-ray diffraction and X-ray reflectivity and diffuse scattering measurements.

Determination of the photocatalytic degradation of methylene blue

The photocatalytic efficiency of the coatings was evaluated acquiring UV-Vis absorption spectra of an aqueous MB solution in time. The UV-Vis absorption spectra of the MB solution were measured in the spectral range 200-800 nm by using a VARIAN Cary 5 spectrophotometer. The light UV-Vis source was a xenon lamp (OSRAM HLX 64610, 50W-12V) inserted in a thin cylindrical container of silica to avoid the contact of the lamp with the aqueous solution. Owing to the presence of the cylindrical container the lamp intensity was attenuated of a factor of 5.

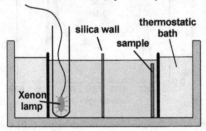

Figure 1 – Scheme of the set-up for the photocatalytic degradation experiments with the thermostatic bath, Xe lamp and sample immersed in the aqueous solution.

DISCUSSION

Here, we discuss in particular the experimental results of two titania films deposited on glass substrates by using the Sol-1 and Sol-2 precursor solutions, respectively. The surface morphology (porosity) is shown in Fig.2 exhibiting the FE-SEM micrographs. The physical parameters are summarized in Table I.

Prior to the experiments of the photocatalytic efficiency of the two titania samples, we first investigated the degradation of the MB due to the action of our UV-Vis lamp. Figure 3 shows the absorbance curves of a 10^{-5} M MB solution contained in a beaker immersed in a thermostatic bath at 20°C and containing the UV-Vis lamp.

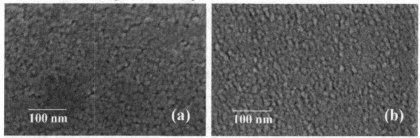

Figure 2 – FE-SEM images of TiO_2 films deposited on soda lime glass substrates by using the precursor solutions Sol-1 (a) and Sol-2 (b), respectively.

The spectra were obtained by sampling 2ml of the MB solution before and after the lamp was switched on for 1h, 1.5h and 3.5h. It is evident that the variation of the absorption peaks is well observed and is time-dependent, clearly demonstrating that a relevant degradation of the methylene blue is induced by the UV-Vis radiation emitted from the xenon lamp.

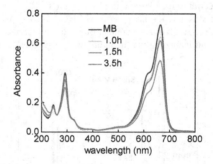

Figure 3 – Absorbance curves of methylene blue solution before (MB curve) and after the Xe lamp was switched on for 1h, 1.5h and 3.5h.

In order to investigate the performance of the two titania coatings on the photocatalytic oxidation of a solution of methylene blue eliminating the BM degradation induced by the xenon lamp, the experimental setup illustrated in Fig.1 was realized. It consists of a rectangular glass tank, immersed in a thermostatic bath at 20°C, and divided into two parts by a wall of silica. On the left side of the tank was put the xenon UV-Vis lamp, whereas on the right side were placed the titania samples. The two sides of the glass tank were filled with a 10^{-5} M solution of MB in pure water. The MB in the left side absorbed the UV-Vis wavelengths that are responsible for the MB degradation so that they did not reach the right side of the tank. The catalytic efficiencies of the titania coatings were then evaluated by monitoring the MB solution contained in the right side of the tank for about 10 hours. More precisely, 2ml of the Mb solution were taken with a pipette before and after the lamp was switched on and analyzed with the spectrophotometer. Before starting the measurements in presence of the two titania samples, the evolution in time of the MB solution in the right side in absence of titania layers was investigated. In this case, no variation in the peak intensity of the absorbance spectra of the MB solution acquired also after 10hrs of UV-Vis irradiation was found.

In Fig.4 the time dependence of the optical absorption of the MB solution of the right side of the tank in presence of titania coating grown by the nanoparticulate bath (a) and the polymeric sol (b) is shown. In Fig.4a only the absorption spectrum acquired after 9.5 hours of exposure to UV-Vis radiation shows a reduction of the peaks intensity. Here, a lowering of about 9.5% of the peak intensity at 663nm is observed indicating that the titania coating exhibits a photocatalytic activity. In Fig.4b a more pronounced lowering of the peak intensities of the absorption spectrum is already observed after 5 hours of irradiation of the sample with the xenon lamp. More precisely, a lowering of about 15.7% (5h) 18.7% (8.5h) and 28.6% (12h) of the peak intensity at 663nm is measured.

This finding shows that both samples exhibit a photocatalytic effect, but the film grown from the polymeric sol exhibits a higher photocatalytic performance than that one grown by the nanoparticulate bath. Since both the photocatalytic and the hydrophilic properties of the titania layers depend upon microstructure and surface structures, especially roughness [7], shape and dimension of the pores [8-9], it is reasonable to suppose that the different photocatalytic performance of the two samples as well as the already observed difference between their hydrophilic characteristic [5] are imputable to the great difference in the morphology and microstructure between the two samples too.

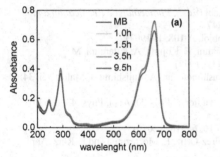 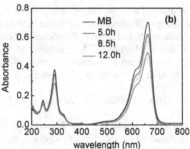

Figure 4 – Optical absorption spectra of the methylene blue (MB) solution in the presence of a sample grown by the nanoparticulate bath (a) and the polymeric sol (b) measured after different UV-Vis exposure time.

In particular, on the basis of the data reported in literature, the titania photocatalytic activity increases with the presence of Ti^{3+} sites on the grains surface [10-11], increasing pore size and film surface area [12.]. So, it is reasonable to suppose that the observed higher photocatalytic performance of the sample grown from the polymeric sol is due to its higher porosity, higher surface area and, consequently, higher number Ti^{3+} sites exposed to the air. It is an interesting result because it shows that also samples having high energy gap can exhibit elevated photocatalytic performance if they, owing to their morphological/microstructural properties, are characterised by large presence of Ti^{3+} active sites.

CONCLUSIONS

Nanocrystalline titania films were prepared by sol-gel synthesis. Our experimental results clearly indicate that the sol - composition and substrate type remarkably influence the microstructural/morphological properties of the titanium dioxide. Consequently, these properties modify the optical response and the hydrophilic performances of the samples. Here, in particular the photocatalytic oxidations of the methylene blue in water of the samples grown on the glass substrate were monitored in order to investigate the influence of the sol-gel precursor solution on the photocatalytic activity of the titania coatings. From our results we can reasonably conclude that the observed higher photocatalytic performance of the sample grown from the polymeric sol is due to its higher porosity, higher surface area and, therefore, higher number Ti^{3+} sites exposed to the air responsible also for the high hydrophilic properties of the sample.

ACKNOWLEDGMENTS

This work is supported by the Regione Puglia (Bari, Italy) – Project HICOGI (Development of materials and processes for the realization of highly innovative coatings for the glass and ophthalmic industry).

REFERENCES

1. T. Watanabe, et al. in: D.E. Olis and H. Al-Ekabi (Eds.), *Photocatalytic Purification and Treatment of Water and Air*, Elsevier (1993) 747.
2. A. Mills and S. LeHunte, J. Photochem. Photobiol. A **108**, 1 (1997).
3. R. Wang, K. Hashimoto, A. Fujishima, M. Chikuni, E. Kojima, A. Kitamura, M. Shimohigoshi and T. Watanabe, Nature **388**, 431 (1997).
4. M. Machida, K. Norimoto, T. Watanabe, K. Hashimoto and A. Fujishima, J. Mater. Sci. **34**, 2569 (1999).
5. M. C. Ferrara, L. Pilloni, S. Mazzarelli, and L. Tapfer, J. Phys. D, Appl. Phys. 43, 095301 (2010).
6. J. Yao and C. Wang, Int. J. of Photoenergy 2010, 643182 (2010).
7. J. Medina-Valtierra, C. Frausto-Reyes, J. Ramirez-Ortiz, E. Moctezuma and F. Ruiz, Appl. Catal. B **76**, 264 (2007).
8. M.A. Carreon, S.Y. Choi, M. Mamak, N. Copra and G.A. Ozin, J. Mater. Chem. **17**, 82 (2007).
9. X. Wang, J.C. Yu, C. Ho, Y. Hou and X. Fu, Langmuir **21**, 2552 (2005)
10. P. Supphasrirongjaroen, W. Kongsuebchart, J. Panpranot, O. Mekasuwandumrong, C. Satayaprasert and P. Praserthdam, Ind. Eng. Chem. Res. **47**, 693 (2008).
11. N. Sakai, A. Fujishima, T. Watanabe, K. Hashimoto. J. Phys. Chem. B **105**, 3023 (2001); U. Diebold, Surf. Sci. Rep. **48**, 53 (2003).
12. X. Wang, J.C. Yu, C. Ho, Y. Hou and X. Fu, Langmuir **21**, 2552 (2005).

Mater. Res. Soc. Symp. Proc. Vol. 1352 © 2011 Materials Research Society
DOI: 10.1557/opl.2011.757

Nb-substituted TiO2 Nanosheet Exfoliated from Layered Titanate for Photocatalysis

Haiyan Song*, Anja O. Sjåstad, Helmer Fjellvåg

Department of Chemistry/SMN, University of Oslo, N-0315 Oslo, Norway

ABSTRACT

Photoelectrochemical behavior and photocatalytic decomposition of Methylene Blue (MB) were studied on $(Nb,Ti)O_2$ nanosheet electrodes and $(Nb,Ti)O_2$ particles produced from nanosheets. We observe that 1% Nb-substitution can drastically increase the photocurrent and photocatalytic activity of $Ti_{1-y}Nb_yO_2$ due to the formation of new defects and electron traps that can promote the separation of photo induced holes and electrons. However, high concentration of electron traps produced by significant Nb-substitution appear to serve as efficient recombination centers that cause loss of photocatalytic activity of the samples.

INTRODUCTION

Titania (TiO_2) has many application areas, including photocatalysis [1], pigments and sensors [2]. Recently, TiO_2 has also offered potentials as a mother compound of electronic materials, such as resistive memories [3], diluted magnetic semiconductors [4], and transparent conductors [5]. In particular, Nb- or Ta-substituted anatase TiO_2 have attracted much attention as promising candidates for practical transparent electrode materials [6-10], which are interesting in e. g., photoelectrochemical systems. Mattsson and co-workers [11] reported photocatalytic decomposition of acetone on TiO_2 and Nb-substituted TiO_2 thin films, and they found that heavy Nb-substitution ($\geq 10\%$) cannot improve the photocatalytic activity of TiO_2, since charge transfer to the conduction band, defect and vacancy site formation must be tuned appropriately.

Recently we reported a novel sol-gel assisted solid-state reaction (SASSR) route for production of phase-pure layered Nb-substituted titanates $Cs_{0.70}Ti_{1.825-x}Nb_xO_4$ [12], in which the Ti^{IV} is partially substituted by Nb^V in the host layers. The SASSR route enables synthesis of Nb-substituted titanates for subsequent exfoliation and opens hence up for design of Nb-substituted TiO_2 films and particles. This soft chemistry route provides products with significant differences with respect to morphology, substitution level and homogeneity, as well extending the thermal stability window of the metastable anatase. Through a systematic study we are presently able to delaminate, reconstruct, and subsequently heat treating $(Nb,Ti)O_2$ powders and films to tune both Nb-substitution as well as TiO_2 polytype [13].

In the present study, we explore the potential of using Nb-substituted TiO_2 nanosheets exfoliated from layered Nb-substituted titanates and transformed into anatase powders and films as the active photocatalytic material through photoelectrochemical behavior of a $(Nb,Ti)O_2$ electrode and photocatalytic decomposition of Methylene Blue (MB) on $(Nb,Ti)O_2$ particles.

EXPERIMENT AND CHARACTERIZATION

$(Nb,Ti)O_2$ nanosheets were prepared as described in [12]. The stable suspensions of exfoliated Nb-titanates were restacked by freeze-drying; yielding a voluminous white solid with a cotton-like appearance (termed restacked material). The dense materials obtained from the nanosheets

---

*Corresponding author: Tel.: +47 22855584; E-mail: haiyans@smn.uio.no

derived from the layered $Cs_{0.70}Ti_{1.825-x}Nb_x\square_{0.175}O_4$ are denoted as $Ti_{1-y}Nb_yO_2$ for convenience in the present paper. TiO_2 nano particles were prepared by a sol-gel process referred to [14]. The samples used for photocatalytic degradation of MB were heated in air at 600 °C for 5 h. Phase purity of the samples were assured from powder X-ray diffraction (Siemens D5000 powder diffractometer, Cu $K\alpha_1$ radiation) , and the morphology of the nanosheets were controlled with AFM (Nanoscope E multimode AFM, Digital Instruments).

Photoelectrochemical behavior of $Ti_{1-y}Nb_yO_2$ was measured with a standard three-electrode system that included an SCE reference electrode and a Pt counter electrode connected to a Parastat 2263 potentiostat. Preparation of $Ti_{1-y}Nb_yO_2$ films (sample electrode) is described in [13]. A sample plate (10×10 mm^2) was used to make an electrode with working area of approximately 20 mm^2. The 1.0 M Na_2SO_4 electrolyte solution was purged with nitrogen prior to the measurements. Irradiation of the sample electrode was carried out with a Solar Simulator (1000 W).

Photocatalytic degradation of MB by $Ti_{1-y}Nb_yO_2$ particles was carried out under irradiation of a 15 W UV lamp (wavelength: 365 nm) positioned over the vessel. Reaction suspensions were prepared by adding 0.05 g material into a 200 ml aqueous MB solution with an initial concentration of 10 mg/L. Prior to photo-degradation, the suspension was magnetically stirred under dark conditions for 1 h to establish an adsorption-desorption equilibrium condition. The aqueous suspension containing MB and photocatalyst was irradiated under the UV light with constant stirring. At given time intervals, analytical samples were taken from the suspension and immediately centrifuge at 9500 rps for 15 min. The filtrate was thereafter analyzed by a UV-vis spectrophotometer (Shimadzu UV-3600) in the range of 190-900 nm.

RESULTS AND DISCUSSION

Nanosheets obtained from the layered titanates

An AFM image of the exfoliated titanate TBA-$Ti_{1.825-x}Nb_xO_4$ (x = 0.02) is shown in Fig. 1. The individual exfoliated nanosheets have an average thickness of 1.0 nm and show no sign of stacking or aggregation, which is in accordance with previous findings [13].

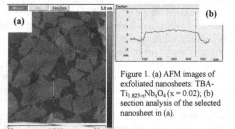

Figure 1. (a) AFM images of exfoliated nanosheets: TBA-$Ti_{1.825-x}Nb_xO_4$ (x = 0.02); (b) section analysis of the selected nanosheet in (a).

Conversion of the nanosheets into dense $Ti(Nb)O_2$ anatase were achieved by freeze-drying of the colloidal TBA-titanates, calcination and heat treatment in air at 600°C. Slightly expanded unit cells of the anatase $Ti_{1-y}Nb_yO_2$ were observed, in agreement with previous findings [13]. This proves that Nb enters as a solid solution substituent and that the soft chemical approach provides products with niobium finely distributed in the TiO_2 matrix.

Photoelectrochemical behavior and photocatalytic activity of $Ti_{1-y}Nb_yO_2$

Figure 2 illustrates the stationary photocurrent-voltage dependencies for the set of samples examined. As evident from the experimental curves, the photocurrent was scarcely observed for the TiO_2 electrode. 1%-Nb substitution causes a drastic increase of the photocurrent, which is reversely decayed when the Nb-substitution level further increases. Miyagi and co-workers [15] showed that titania substituted with Nb leads to the formation of new defects and electron traps, which can promote the separation of photo-induced holes and electrons. But high concentration of electron traps can be assumed to serve as efficient recombination centers that caused complete loss of photocatalytic activity of the modified titania samples.

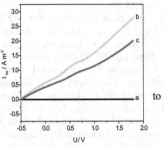

Figure 2. Experimental dependencies of stationary photocurrent on applied voltage for $Ti_{1-y}Nb_yO_2$ grown on Si (111) electrodes: (a) y = 0; (b) y = 0.01 and (c) y = 0.015.

The photocatalytic bleaching of the organic dye MB was measured to examine the activity in more detail.

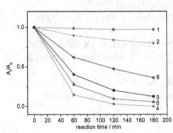

Figure 3. Photocatalytic decomposition of MB under dark (1), UV light without catalyst (2) and in the presence of $Ti_{1-y}Nb_yO_2$ from nanosheets (3) y= 0, (4) y= 0.01, (5) y= 0.015. (6) is TiO_2 nanoparticles prepared by sol-gel process. All the catalysts were heated at 600 °C for 5h.

Figure 3 shows the changes in peak absorbance at 650 nm of MB versus UV irradiation time. The constant decrease in optical absorption of MB indicates the photocatalytic decomposition of MB by $Ti_{1-y}Nb_yO_2$ formed from nanosheets. Nb-substitution can enhance the photocatalytic activity of $Ti(Nb)O_2$ significantly, but when the Nb content exceeds 1 %, the activity decreases reversely, which is in accordance with the photoelectrochemical behavior of $Ti_{1-y}Nb_yO_2$ films above. It implies that there is an optimum molar content of Nb = 1% for substitutued TiO_2 nanosheets used for photocatalyst. To compare the photocatalytic activity of $Ti_{1-y}Nb_yO_2$ and titania prepared by traditional synthesis process, TiO_2 nano particles synthesized by sol-gel process was also tested (Fig. 3, sample labeled (6)). It clearly appears that titania produced from nanosheets shows higher activity than that produced by sol-gel process, maybe due to its larger specific surface area.

CONCLUSIONS

Single-layered Nb-substituted titanate nanosheets obtained by exfoliating TBA-intercalated Nb-substituted titanates transformed into anatase $Ti_{1-y}Nb_yO_2$ powders and films where tested for photocatalytic activity. Results show that 1% Nb-substitution drastically increases the photocatalytic activity due to the formation of new defects and electron traps that can promote the separation of photo-induced holes and electrons. However, higher Nb-substitution levels caused loss of photocatalytic activity of the samples.

ACKNOWLEGDEMENT

The authors acknowledge financial support from the Research Council of Norway (NANOMAT, Grant No.163565 431).

REFERENCES

1. A. Fujishima and K. Honda, *Nature* **238**, 37(1972).
2. O. Carp, C. L. Huisman, and A. Reller, *Prog. Solid State Chem.* **32**, 33 (2004).
3. M. Fujimoto, H. Koyama, M. Konagai, Y. Hosoi, K. Ishihara, S. Ohnishi, and N. Awaya, *Appl. Phys. Lett.* **89**, 223509 (2006).
4. Y. Matsumoto, M. Murakami, T. Shono, T. Hasegawa, T. Fukumura, M. Kawasaki, P. Ahmet, T. Chikyow, S. Koshihara, and H. Koinuma, *Science* **291**, 854 (2001).
5. Y. Furubayashi, H. Hitosugi, Y. Yamamoto, K. Inaba, G. Kinoda, Y. Hirose, T. Shimada, and T. Hasegawa, *Appl. Phys. Lett.* **86**, 252101 (2005).
6. T. Hitosugi, Y. Furubayashi, A. Ueda, K. Itabashi, K. Inaba, Y. Hirose, G. Kinoda, Y. Yamamoto, T. Shimada, and T. Hasegawa, *Jpn. J. Appl. Phys.* **44**, L1063 (2005).
7. D. Kurita, S. Ohta, K. Sugiura, H. Ohta, and K. Koumoto, *J. Appl. Phys.* **100**, 096105 (2006).
8. S. X. Zhang, S. Dhar, W. Yu, H. Xu, S. B. Ogale, and T. Venkatesan, *Appl. Phys. Lett.* **91**, 112113 (2007).
9. M. A. Gillispie, M. F. A. M. van Hest, M. S. Dabney, J. D. Perkins, and D. S. Ginley, *J. Appl. Phys.* **101**, 033125 (2007).
10. J. Osorio-Guillen, S. Lany, and A. Zunger, *Phys. Rev. Lett.* **100**, 036601 (2008).
11. A. Mattsson, M. Leideborg, K. Larsson, G. Westin and L. Osterlund, *J. Phys. Chem. B* **110**, 1210 (2006).
12. H. Y. Song, A. O. Sjåstad, Ø. B. Vistad, T. Gao and P. Norby, *Inorg. Chem.* **48**, 6952 (2009).
13. H. Y. Song, A. O. Sjåstad, H. Fjellvåg, H. Okamoto, Ø. B. Vistad, B. Arstad and P. Norby, *J. Solid State Chemistry*, submitted, (2011).
14. M. H. Lee, Y. H. Park and C. K. Yang, *J. American Ceramic Society* **70**, C35 (1987).
15. T. Miyagi, M. Kamei, I. Sakaguchi, T. Mitsuhashi, A. Yamazaki, *Jpn. J. Appl. Phys.* **43**, 775 (2004).

Mater. Res. Soc. Symp. Proc. Vol. 1352 © 2011 Materials Research Society
DOI: 10.1557/opl.2011.874

Rational Designs on TiO$_2$-based Nanocomposites for Solar Photocatalytic Purification

Shanmin Gao[1,2] and Tao Xu[*1]

[1] Department of Chemistry and Biochemistry, Northern Illinois University, DeKalb, IL 60115, U.S.A.

[2] School of Chemistry and Materials Science, Ludong University, Yantai 264025, Shandong, PR China

ABSTRACT

We report an template-free process to fabricate S-C-codoped and (I$_2$)$_n$-C-codoped meso/nanoporous TiO$_2$ nanocrystallites. Methylene blue solutions are used as a model pollute to evaluate the sorption and photocatalytic activity of the samples under visible light radiation. The high photocatalytic activity in visible light region of our samples is attributed to numerous oxygen vacancies, large specific surface area and the continuous states in the band gap of TiO$_2$ introduced by I$_2$ or S doping.

*T.X. Email: txu@niu.edu

INTRODUCTION

Semiconductor-catalyzed photo-oxidation powered by sunlight is a promising and cleaner scheme for energy- and environment-sustainable degradation of organic and biological pollutants in wastewater[1,2]. Anatase TiO$_2$ has attracted much attention due to its abundance, high photostability, and environmentally benign properties[3].

Several critical problems, however, impede the efficiency of TiO$_2$-based photocatalysts for wastewater treatment. (1) The large band gap of pure TiO$_2$ (~3.2 eV) only slightly overlaps with the solar spectrum in the UV regions, so, it is crucial to enhance the light harvesting efficiency of TiO$_2$ in the visible region[4,5]. (2) The photo-induced electron-hole pairs can either be trapped or recombine at the defect states inside the bulk of TiO$_2$ structure before reaching the external surface[6,7]. (3) The conflicting requirements of particle size. Small particles possess more surface area. However, reduction in particle size introduces more internal crystal defects and particle-particle interfaces that act as charge trap states to diminish the effective transport of photoinduced charges to the external surface sites[8,9]. (4) The constraint in the effective mass transport and the subsequent surface adsorption of the organic species in the photocatalyst matrix[10].

For all these reasons, it is highly desirable to design anatase TiO$_2$-based photocatalysts with visible light-driven photoactivity, strategically designed structures for large attainable surface area, fast charge transport to surface sites and facile mass flow channels for molecular transport, and surface engineering for enhanced adsorption of organic pollutant molecules.

EXPERIMENTAL DETAILS

The preparation, characterization and photocatalytic activity measurements have been described elsewhere[11,12]. The samples after heat treatment were denoted as STC-T for S-C-codoped TiO_2 samples and PIT-T for $(I_2)_n$-C-codoped TiO_2 samples, where the second T refers to temperature used for the heat treatment. For comparison, C-doped TiO_2 was also prepared using the same method under calcination temperature of 400°C(denoted as TC-400), except that no sulfur or I_2 hydrosol was used.

DISCUSSION

S-C-codoped TiO_2 and $(I_2)_n$-C-codoped TiO_2 was prepared via the hydrolysis of TBOT in S or I_2 hydrosol without any addition of templates or surfactants, followed by a heat treatment in air at elevated temperatures for 3 h. The samples were characterized by TEM as shown in Figure 1. Each individual sphere is composed of a large number of much smaller and loosely packed TiO_2 nanoparticles. As a result, the interstitial voids among these primary nanoparticles constitute a short-range disordered nanoporous structure, which is termed as primary nanopores. In comparison, the voids among the large spheres (termed as secondary mesoparticles) also create pores. The morphology of the primary pores gradually changes as the size of the primary nanoparticles increases with the increasing temperature of heat treatment.

Figure 2 shows the XRD patterns of the samples heat treatment at 400°C and 600°C. The pattern showed that all samples have anatase TiO_2 structure. As the heat treatment temperature increases, the XRD peaks become sharper and more intense due to the formation of larger grains as summarized in Table 1. Since the heat treatment is a necessary step for doping and for the formation of bimodal meso/nanoporous structure, the high thermal stability of our samples is a favorable feature for maintaining anatase under high temperature treatment[13].

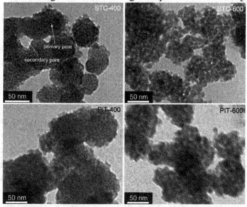

Figure 1. TEM images of STC and PIT calcined for 3 h at 400°C and 600°C.

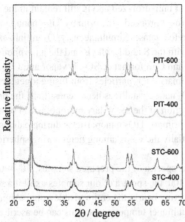

Figure 2. XRD patterns of STC and PIT calcined for 3 h at 400°C and 600°C.

Table 1. Physicochemical properties of the samples from XRD and N_2 sorption analysis.

Sample	S_{BET} (m^2/g)	Pore volume (cm^3/g)	Average pore size (nm)	Crystalline size (nm)
STC-400	249.33	0.4918	7.8745	4.2
STC-600	144.14	0.4612	9.4631	8.3
PIT-400	186.6	0.343	8.586	6.1
PIT-600	118.2	0.309	5.249	7.8

The pore size and surface area of the samples are further characterized by nitrogen adsorption–desorption isotherm measurements. Table 1 summarized the physical properties of the samples obtained after heat treatment at various temperatures. These meso/nanoporous structures with high specific surface area are of particular interest, since they can provide more active sites to adsorb pollute molecules and enhance catalytic activity [11,12].

XPS analysis has been used to confirm the presence, contents and chemical states of S, I_2, C and Ti in the samples[11,12]. The XPS survey spectra of the samples calcined at 400°C indicating the presence of Ti, O, C, S or I. For the S_{2p} peak, sulfur is present in different oxidation states depending on the calcination temperature. The XPS spectrum of the I 3d region shows doublet peaks at 620.5 eV (I 3d3/2) and 630.5 eV (I 3d5/2), which is equivalent to those in molecular I_2.

We propose a possible formation mechanism of the hierarchically meso/nanoporous TiO$_2$ spheres. We think that the S or I_2 hydrosol plays an important role in the formation of these porous structures.

First, the addition of TBOT in the S or I_2 hydrosol at room temperature leads to the nucleation of primary amorphous TiO$_2$ nanoparticles on the surface of the S or I_2 particles. The rate of hydrolysis is relatively low, thus the hydrolysis cannot complete at room temperature. As a consequence, a large amount of unhydrolyzed alkyls still remain in the xerogel powders.

As the calcination of the xerogel proceeds, the primary TiO_2 nanoparticles undergo a phase transition from amorphous form to anatase. Simultaneously, O_2 initializes a set of oxidation–reduction reactions with the S and I_2, alkyls and the amorphous TiO_2, giving rise to the partial vaporization of S, I_2 and C in format of SO_2, I_2 vapor and CO_2. The escape of these gaseous species prevents the primary TiO_2 particles from fusing together so that the interstitial voids among these primary TiO_2 nanocrystallines hence constitutes the primary nanopores.

On the other hand, the formation of the secondary mesopores can be attributed to the aggregation of the neighboring primary TiO_2 nanoparticles during calcinations[14]. The secondary mesopores are essentially the voids among neighboring spherical jumbles of the primary TiO_2 nanocrystallines.

Figure 3 collects the UV-Vis spectra of the prepared samples, TC-400 and P25. In comparison to the P25 and TC-400, the spectra of our samples exhibit a strong broad absorption band between 400–800 nm, covering nearly the entire visible range. Apparently, the absorbance decreases for sample prepared at higher temperatures. This can be ascribed to the additional loss of S or I_2 by sublimation at higher temperatures. Nonetheless, the absorption in the visible-light region implies that the prepared co-doped samples can be activated by visible light and that more photogenerated electrons and holes can be created and they can participate in the photocatalytic oxidation reactions.

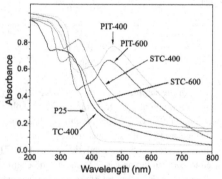

Figure 3. UV-vis DRS of TC-400, P25, STC-T and PIT-T.

The decomposition profiles of methylene blue(MB) are collected in Figure 4. The concentration of MB decreases slowly for all TiO_2 samples analyzed before the light is turned on. However, it is still noticeable that the concentrations of MB in solution containing SCT-T and PIT-T samples are lower than those in the presence of P25 and TC-400. This trend agrees with the fact that the meso/nanoporous samples possess more specific surface area, thus adsorb more MB molecules and cause more depletion of methylene blue in solution phase.

Compared to the only TC-TiO₂ and P25, significantly enhanced photocatalytic oxidation activity on the decomposition of MB under visible light is observed for meso/nanoporous samples. The MB solutions containing the STC-T and PIT-T samples were almost fully degraded and became transparent within about 30 min. It should be noted that STC-400 sample showed the highest photocatalytic oxidation on MB.

The photocatalytic activity of TC-400 approached to that of P25. This might be ascribed to the following facts that, on the one hand, at low temperature, the TC-400 sample prepared by conventional hydrolysis method was lower crystalline, while the P25 had high crystalline, on the other hand, the TC-400 had the minimum specific surface area, pore volume and average pore size, etc. The photocatalytic activity of TiO_2 is known to be a function of its surface properties and to increase with increase in the surface area of the sample used [15]. Therefore, the TC-400 sample showed a low photocatalytic activity.

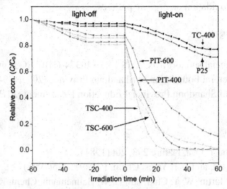

Figure 4. The photodegradation of methylene blue solutions by using STC-T, PIT-T, P25 and TC-400 as photocatalysts under visible light irradiation.

The enhanced photocatalytic activity of our samples is a product of several factors. First, the doping of S or $(I_2)_n$ introduces continuous states residing in between the VB and CB of TiO_2[16]. We believe these S or I-induced continuous states can either accept visible light-excited electrons from the VB of TiO_2, and/or provide visible-light excited electrons from these S or I_2-induced states to the CB of TiO_2. Second, the carbon doping enhances the conductivity of TiO_2, so that the photo-induced hot carriers can rapidly transfer to the surface region and participate in the desired photo-oxidation reactions[11,12]. Third, the highly crystalline anatase phase also promotes the transfer of photoelectrons from bulk to surface, thus inhibits charge recombination in the bulk of TiO_2[2]. Because the photo-degradation occurs at the surface of TiO_2, the organic pollute molecules must be pre-concentrated at the TiO_2 surface in order to react with the trap hot carriers and the reactive radicals. Thus, the carbon doping and the increased surface area of the meso/nanoporous TiO_2 together enhance the adsorption of MB molecules onto the surface of TiO_2 particles. This is evidenced by the pronounced more and faster deletion of methylene blue molecules in solution under darkness (Figure 4) in comparison to P25 and TC-400 sample. In addition, large surface area can accommodate more surface-adsorbed water and hydroxyl groups that act as photoexcited hole traps and produce hydroxyl radicals for the degradation of organic pollute molecules[17,18].

CONCLUSIONS

Bimodal meso/nanoporous S-C-codoped and $(I_2)_n$-C-codoped TiO_2 with high crystallinity and high surface area has been synthesized through a simple route. The resulting meso/nanoporous TiO_2 shows enhanced visible light-driven photocatalytic oxidation on methylene blue, which is attributed to the high specific surface area of the meso/nanoporous TiO_2 structure, the high crystallinity of anatase TiO_2, the enhanced electrical conductivity due to carbon-doping. The architecturally controllable design in the morphology and doping region of the TiO_2 nanostructure presented in this work provide new basic science to further enhance visible light-driven photo-oxidation on organic pollutes.

ACKNOWLEDGMENTS

TX acknowledges the financial support from ACS-PRF (46374-G10). SG is grateful to the financial support from Doctoral Foundation of Shandong Province (2007BS04040) and Scientific Research Fund of Shandong Provincial Education Department (J07WA04).

REFERENCES

1. M. Fujihira, Y. Satoh, and T. Osa, Nature 293, 206(1981).
2. A. L. Linsebigler, G. Q. Lu, and J. T. Yates, Chem. Rev. 95, 735(1995).
3. M. R. Hoffmann, S. T. Martin, W. Y. Choi, and D. W. Bahnemann, Chem. Rev. 95, 69(1995).
4. E. Finazzi, C. Di Valentin, and G. Pacchioni, J. Phys. Chem. C 113, 220(2009).
5. N. O. Gopal, H. H. Lo, and S. C. Ke, J. Am. Chem. Soc. 130, 2760(2008).
6. T. Tachikawa, M. Fujitsuka, and T. Majima, J. Phys. Chem. C 111, 5259(2007).
7. N. Serpone, J. Phys. Chem. B 110, 24287(2006).
8. Z. Yang, T. Xu, Y. S. Ito, U. Welp, and W. K. Kwoko, J. Phys. Chem. C 113, 20521(2009).
9. Z. Yang, T. Xu, S. Gao, U. Welp, and W. K. Kwok, J. Phys. Chem. C 114, 19151(2010).
10. N. Papageorgiou, C. Barbé, and M. Grätzel, J. Phys. Chem. B 102, 4156(1998).
11. P. Xu, T. Xu, J. Lu, S. M. Gao, N. S. Hosmane, B. B. Huang, Y. Dai, and Y. B. Wang, Energ. Environ. Sci. 3, 1128(2010).
12. P. Xu, J. Lu, T. Xu, S. Gao, B. Huang, and Y. Dai, J. Phys. Chem. C 114, 9510(2010).
13. M. Janus, M. Inagaki, B. Tryba, M. Toyoda, and A. W. Morawski, Appl. Catal. B: Environ. 63, 272(2006).
14. T. Z. Ren, Z. Y. Yuan, and B. L. Su, Chem. Phys. Lett. 374, 170(2003).
15. J. G. Yu, M. H. Zhou, B. Cheng, H. G. Yu, and X. J. Zhao, J. Mol. Catal. A: Chem. 227, 75(2005).
16. S. Tojo, T. Tachikawa, M. Fujitsuka, and T. Majima, J. Phys. Chem. C 112, 14948(2008).
17. C. P. Sibu, S. R. Kumar, P. Mukundan, and K. G. K. Warrier, Chem. Mater. 14, 2876(2002).
18. V. Subramanian, E. E. Wolf, and P. V. Kamat, J. Am. Chem. Soc. 126, 4943(2004).

Mater. Res. Soc. Symp. Proc. Vol. 1352 © 2011 Materials Research Society
DOI: 10.1557/opl.2011.782

Fabrication of Rh-doped TiO₂ nanofibers for Visible Light Degradation of Rhodamine B

Emilly A. Obuya[1], William Harrigan[1], Tim O'Brien[1], Dickson Andala[2], Eliud Mushibe[1] & Wayne E. Jones Jr[1*].

[1]Department of Chemistry, State University of New York at Binghamton, 4400 Vestal Parkway East, Vestal, NY 13902, USA.
[2]Department of Chemical Engineering, Pennsylvania State University, 25 Fenske, University Park, PA 16802, USA.

*Author to whom correspondence should be addressed; e-mail; wjones@binghamton.edu; Tel: +1-607-777-2421; Fax: +1-607-777-4478.

ABSTRACT

The synthesis and application of environmentally benign, efficient and low cost heterogeneous catalysts is increasingly important for affordable and clean chemical technologies. Nanomaterials have been proposed to have new and exciting properties relative to their bulk counterparts due to the quantum level interactions that exist at nanoscale. These materials also offer enormous surface to volume ratios that would be invaluable in heterogeneous catalysis. Recent studies point at titanium dioxide nanomaterials as having strong potential to be applied in heterogeneous photocatalysis for environmental remediation and pollution control. This work reports the use of surface modified anatase TiO₂ nanofibers with rhodium (Rh) nanoparticles in the photodegradation of rhodamine B (RH-B), an organic pollutant. The dimensions of TiO₂ nanofibers were 150±50 nm in diameter and the size of the Rh nanoparticles was ~5 nm. The Rh-doped TiO₂ catalyst exhibited an enhanced photocatalytic activity in photodegradation of rhodamine B under visible light irradiation, with 95 % degradation within 180 minutes reaction time. Undoped TiO₂ did not show any notable phocatalytic activity under visible light.

INTRODUCTION

TiO₂ nanomaterials are used as supports for heterogeneous nucleation of metal nanoparticles, with a strong potential as photocatalysts in environmental remediation. TiO₂ has band gap energies of 3.0 - 3.2eV,[1] that require UV light to generate excitons involved in the photodegradation process. Efforts to shift the optical response of titania to the visible region have been actively studied for a number of years[2],[3],[4]. Doping TiO₂ has been established as a way of shifting the absorption maximum band to visible range by introducing electronic states within the band gap[5]. These intermediate states trap the photogenerated electron-hole pairs while providing a lower energy path for excitation. Additionally, research has shown that a shift of titania's absorption band to visible range can also be induced by introduction of crystal defects, which can be created through phase transformation[6] and/or variation of the crystal size of the titania[7]. The inherent differences leading to the red shift in these cases is that phase transformation and crystal size decreases/increases the band gap of titania whereas doping creates energy states within the band gap[6].

We are exploring the design, synthesis and characterization of novel TiO₂ based nanomaterials for sustainable environmental remediation as well as greener chemical transformations. This study will highlight the modification of electronic band structure of TiO_2 nanofibers by photoinduced nucleation of Rh nanoparticles. The highly folded and mesoporous surface of electrospun TiO_2 provided an effective support for seeding, nucleation and growth of metal nanoparticles without the need for external stabilizing agents. The doped TiO_2 was used to harvest visible light for the formation of electron-hole pairs that generated reactive radicals on the TiO_2 surface for the degradation of model organic dyes.

EXPERIMENTAL

Materials

Polymethymethacrylate (PMMA), Mw 350,000, titanium isopropoxide (TiP), dichloromethane, rhodium(III) chloride ($RhCl_3$), ethylene glycol (EG), polyvinyl pyrrolidone (PVP), Mw 55 000, N,N-dimethylformamide (DMF), rhodamine B (RH-B), were all purchased from Sigma Aldrich and used as received.

Instrumental

The morphology and size of the TiO_2 nanofibers were observed using a Hitachi S-570 or Supra Zeiss 55VP SEM (Scanning electron Microscope) Model. The surface area, pore volume and pore size distribution were measured using the Brunauer-Emmett-Teller (BET) method on Micrometrics ASAP 2020 at −196 °C. Samples for BET were first heated at 250 °C to remove adsorbed water, followed by degassing at this temperature for 4 hours prior to analysis. TEM images were obtained on a JEOL 2010 FETEM instrument. The samples were dispersed in ethanol by sonication, then drop-cast onto a lacey carbon-coated Cu grid. Uv-vis analysis of the aliquots was done on a 8452A Hewlett Packard Diode Array spectrophotometer instrument from 190 to 820 nm range. Uv-vis sample analyses for the catalysts were performed in methanol, and distilled water for the RH-B solutions.

Electrospinning of TiO_2 nanofibers

A sol-gel technique was used in the preparation of the TiO_2 nanofibers from a solution of titanium isopropoxide[8]. A PMMA:TiP composite solution (in dichloromethane) was continuously stirred at room temperature to attain a suitable viscosity for the electrospinning process. Into this solution, DMF was added to ensure a high dielectric constant to withstand the high electric voltages during electrospinning. A high voltage power supply of 25 kV was adequate to induce jet instability that initiated an elongation process allowing the TiP to be collected as long and thin fibers. These fibers were then exposed to moisture overnight to completely hydrolyze the TiP to $Ti(OH)_4$. This was followed by calcination at 500 °C for approximately 4 hours to remove the PMMA component giving TiO_2 nanofibers, Figure 2a.

Photodeposition of Rh on TiO_2 nanofibers

A typical procedure included heating of 10 mL of ethylene glycol at 110 °C for 30 min, followed by addition of TiO_2 fibers for another 30 min with UV light irradiation. Subsequently, $RhCl_3$ and PVP (1:20 mole ratio) were simultaneously added to the ethylene glycol solution at 110 °C and left to react for another 1.5 hours. The UV light was turned off on addition of the metal precursor to initiate the nucleation process, which was indicated by the change in color of the reaction mixture from deep red to grey. At the end of the reaction, the catalyst was washed several times with acetone under centrifugation and dried in the oven at 120 °C for 3 hours to remove remnants of PVP and ethylene glycol from the catalyst surface.

Photodegradation of RH-B

About 50 mg of the Rh-TiO$_2$ catalyst was sonicated in 5 mL distilled water to fully disperse the catalyst in solution. The catalyst was then added to a 10^{-5} M solution of RH-B in distilled water and stirred for 30 min in the dark in order to achieve an adsorption/desorption equilibrium of RH-B and the catalyst. The mixture was then placed by the window with direct sunlight exposure for visible light irradiation. For Uv-vis analysis of the degradation process, 0.5 mL aliquots were periodically withdrawn from the reacting solution and centrifuged (in 5 mL distilled water) for 10 min to remove the catalyst particles. The precipitated catalyst was collected from each aliquot, cleaned and dried at 120 °C for recycling.

RESULTS AND DISCUSSION

Material Characterization

Nitrogen adsorption/desorption isotherms of the Rh-TiO$_2$ nanostructured material were recorded against relative pressure of nitrogen gas at 77 K, Figure 1. From the sharp inflection point at a relative pressure of 0.8, a BET surface area of ~288 m^2/g was determined. The hysteresis loop indicated a type IV isotherm suggesting mesoporosity of the TiO$_2$ nanomaterials, with pore condensation at high pressure. Inset, the BJH (Brunauer-Joyner-Halenda) analyses indicate a narrow pore distribution centered at ~12.5 nm, which is further indication of the mesoporous nature of our Rh-TiO$_2$ nanostructured material. The increased porosity of the titania nanofibers makes them a favorable material for heterogeneous catalysis due to their ability to absorb an increased amount of the reacting molecules. The pores also allowed for the effective control of metal nanoparticles sizes by providing suitable seeding sites, Figure 2c.

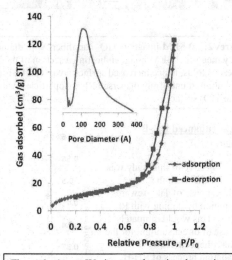

Figure 1. A type IV nitrogen adsorption-desorption isotherm indicative of a highly mesoporous surface for the Rh-TiO$_2$ catalyst. BET surface area of 288 m^2/g was calculated from the sharp inflection point at 0.8 P/P$_0$. Inset is the BJH analyses for pore size distribution of the Rh-TiO$_2$

SEM analysis confirmed that the TiO_2 nanofibers had diameters of 150 ± 50 nm, Figure 2a. The TiO_2 surface was highly folded thus increasing surface area for catalysis as well as availability of numerous nucleation sites for nanoparticle deposition, Figure 2b. The nanofibers exhibited a highly mesoporous surface that is essential for the effective adsorption and free flow of reacting molecules for heterogeneous catalysis. Additionally, TiO_2 nanofibers were used as a support matrix for the in-situ reduction and nucleation of the Rh nanoparticles of ~5 nm, as indicated by TEM image, Figure 2c.

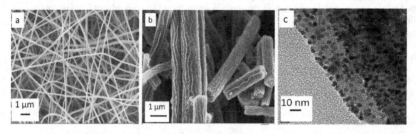

Figures 2: (a) SEM image of TiO_2 nanofibers with diameters of 150 ± 50 nm. **(b)** A highly magnified **SEM** image indicating a section of the TiO_2 nanofiber with numerous folds, hence increased surface area. **(c) TEM** image showing the distribution of uniformly dispersed Rh nanoparticles with average diameters of ~5 nm on the TiO_2 surface.

Visible light-enhanced RH-B degradation

The objective of this study was to obtain an extension of the wavelength response of TiO_2 towards the visible region by doping with Rh nanoparticles. This would ultimately enable the use of direct sunlight for the visible light-enhanced photodegradation process of RH-B. The Rh-doped TiO_2 produced a large bathochromic shift of ~210 nm in the TiO_2 maximum absorption band, suggesting a modification in TiO_2 electronic band structure, Figure 3.

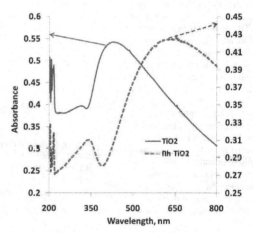

Figure 3. Absorption spectra of the Rh-TiO_2 catalyst compared to undoped TiO_2 with maximum absorption peaks at ~622 nm and

Role of TiO$_2$ in RH-B degradation

Doping of TiO$_2$ with Rh nanoparticles lowered the band gap energy for excitation, thereby allowing photogeneration of conduction band electrons and valence band holes using visible light irradiation. The presence of intermediate electronic states from the Rh nanoparticles further prevented the recombination of the electron-hole pair, thus improving the efficiency of the photocatalyst. The TiO$_2$ crystal structure has oxygen ion vacancies, which together with the terminal hydroxyl groups react with the electron-hole pair to generate highly active hydroxyl and/or superoxide radicals. The valence band holes, together with the photogenerated radicals progressively attack the non-bonded electron pairs on the RH-B functional groups as well as targeting the extended conjugation in structure, Figure 4a. The extent of the photodegradation reaction was monitored by the disappearance of the characteristic absorption peak (~554 nm) of RH-B, Figure 4b.

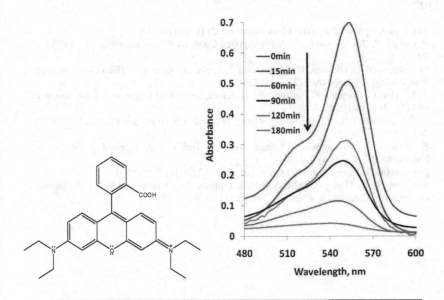

Figures 4: (a) Structure of RH-B. **(b)** Photodegradation of RH-B with Rh-TiO$_2$ under visible light irradiation, showing complete degradation after 180 min reaction time.

CONCLUSION

The controlled nucleation and stabilization of well dispersed ~5 nm Rh nanoparticles onto mesoporous TiO_2 nanofibers was achieved. The nanoporous nature of TiO_2, coupled with its photoactive efficiency allowed for a simple, cheap and sustainable method for the fabrication of highly active photocatalysts without the need for external stabilizing agents. The Rh-TiO_2 catalyst was successful in the visible-light degradation of RH-B, with 95 % of the dye degraded within 180 min reaction time.

ACKNOWLEDGEMENTS

ACS-PRF 46117 as well as SUNY Research Foundation for funding and SUNY Binghamton Chemistry Department for financial support

REFERENCES

1]   M.I. Litter, Applied Catalysis B-Environmental 23 (1999) 89-114.
2]   P. Lianos, P. Bouras, and E. Stathatos, Applied Catalysis B-Environmental 73 (2007) 51-59.
3]   M. Maicu, M.C. Hidalgo, G. Colon, and J.A. Navio, Journal of Photochemistry and Photobiology a-Chemistry 217 (2011) 275-283.
4]   Z.-M. Dai, G. Burgeth, F. Parrino, and H. Kisch, Journal of Organometallic Chemistry 694 (2009) 1049-1054.
5]   Y. Su, Y. Xiao, Y. Li, Y. Du, and Y. Zhang, Materials Chemistry and Physics 126 (2011) 761-768.
6]   P. Bouras, E. Stathatos, P. Lianos, and C. Tsakiroglou, Applied Catalysis B-Environmental 51 (2004) 275-281.
7]   E. Stathatos, T. Petrova, and P. Lianos, Langmuir 17 (2001) 5025-5030.
8]   E.A. Obuya, W. Harrigan, D.M. Andala, J. Lippens, T.C. Keane, and W.E.J. Jr, Journal of Molecular Catalysis A: Chemical 340 (2011) 89-98.

Mater. Res. Soc. Symp. Proc. Vol. 1352 © 2011 Materials Research Society
DOI: 10.1557/opl.2011.760

(Green) Photocatalytic Synthesis Employing Nitroaromatic Compounds.

Ralf Dillert, Amer Hakki, and Detlef W. Bahnemann
Institut für Technische Chemie, Gottfried Wilhelm Leibniz Universität, Callinstrasse 3, 30167
Hannover, Germany.

ABSTRACT

The combination of a solid photocatalyst (TiO_2) and a co-catalyst (p-toluenesulfonic acid) has been successfully applied for the light-induced conversion of nitroarenes in O_2-free ethanolic suspensions yielding substituted quinolines and tetrahydroquinolines, while in the presence of TiO_2 loaded with a noble metal (Pt, Pd) N-alkylarylamines and N,N-dialkylarylamines were formed. Depending on the compounds that have been detected by GC–MS the reaction mechanism is discussed comprising the formation of anilines and ethanal by a photocatalytic reaction step and their subsequent thermal reactions to quinolines, tetrahydroquinolines, and N-alkylated anilines via a Schiff base as an intermediate product.

INTRODUCTION

Organic photosynthesis in the presence of semiconducting metal oxides as heterogeneous photocatalysts has become an important area of research. Experimental results have been summarized in several review articles [1-5].

The light-induced charge separation occurring in TiO_2 particles under irradiation with UV(A) light creates both, a reduction center and an oxidation center at the particles' surface. These active centers can induce the reduction as well as the oxidation of adsorbed organic species by interfacial electron transfer (Scheme 1). In principle, this unique feature allows multistep reactions at the surface of a single photocatalyst particle: an intermediate generated at one type of the active centers can react with the intermediate generated at the other type of reaction center allowing a multi-step synthesis in a one-pot reaction.

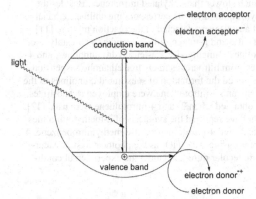

Scheme 1. Basic mechanism of photocatalysis.

It has been shown by several authors that light-induced reactions of nitroarenes in the presence of alcohols and TiO$_2$ as the photocatalyst yield a variety of different products depending on the reaction conditions, i.e., the presence of a co-catalyst and / or metal deposits enabling the electron shuttle at the photocatalyst´s surface.

In an early contribution Mahdavi et al. have reported that an amino compound and an aldehyde are formed in high yields by UV(A) irradiation of a suspension of TiO$_2$ in a primary alcohol containing a nitro compound. [6]. Ferry and Glaze have investigated the mechanism of the photocatalytic reduction of some nitroarenes in suspensions of TiO$_2$ in methanol and 2-propanol in the absence of molecular oxygen in more detail. They have shown that conduction band electrons (free or trapped as Ti(III)) are the principal reductive species driving the reactions of nitro compounds to the corresponding amino compounds. Secondary reductive radicals generated from alcohol oxidation by valence band holes were found to play no role in the formation of the amino compounds [7]. Valenzuela and co-workers have investigated the photocatalytic reactions of nitrobenzene in deaerated non-aqueous suspensions of TiO$_2$ in ethanol. Aniline and ethanal (acetaldehyde) were found to be the main products of the reduction of the nitroarene and the simultaneous oxidation of ethanol. Nitrosobenzene and N-hydroxyaniline were found as the intermediates of the reduction of nitrobenzene by valence band electrons at the TiO$_2$ surface; ethanoic acid (acetic acid) was identified as the oxidation product of ethanal. 1,3-Dimethyl-1H-indole, 2,3-dimethyl-1H-indole, 2-methylquinoline, 2-methyl-quinoline-1-oxide, and 3-methyl-quinoline-2(1H)-one, probably formed via reactions between aniline and the products of the photocatalytic ethanol oxidation, were identified by GC-MS [8]. Valenzuela and co-workers have also demonstrated the synthesis of imines from nitrobenzene and alcohols (primary C1 – C5 alcohols, and 2-propanol) in the presence of UV(A) irradiated TiO$_2$ [9].

Hakki et al. have employed photo-platinized TiO$_2$ nano-sized particles as the photocatalyst for the one-pot synthesis of N-alkylarylamines from nitroarenes and ethanol through the UV(A)-irradiation of Ar-purged suspensions of this catalyst in ethanolic solutions containing nitro aromatic compounds. GC/MS analysis of the irradiated mixtures showed a gradual conversion of the nitroarenes to N-alkylarylamines during the illumination time. The produced N-alkylarylamines were further transformed by prolonged irradiation time into N,N-dialkylarylamines. [10].

Park et al. have synthesized 4-ethoxy-2-methyl-1,2,3,4-tetrahydroquinolines from nitroarenes (nitrobenzene, 3- and 4-nitrotoluene) and ethanol in deaerated, non-aqueous TiO$_2$ suspensions under irradiation with UV(A) light. The yields of the respective products were about 70%. Under the same reaction conditions nitroarenes having oxygen or amino substituents (e.g. 3-methoxy-nitrobenzene) were transformed much slower than alkylated nitroarenes. Besides the respective 4-ethoxy-2-methyl-1,2,3,4-tetrahydroquinolines the corresponding anilines, ethanal, and 1,1-diethoxyethan (acetaldehyde diethyl acetal) were detected in the reaction mixture [11].

Hakki et al. have used a combination of TiO$_2$ and p-toluenesulfonic acid as a co-catalyst for the light-induced synthesis of 2-methylquinolines, 2-methyl-1,2,3,4-tetrahydroquinolines, and 4-ethoxy-2-methyl-1,2,3,4-tetrahydroquinolines from nitroarenes in O$_2$-free ethanolic suspensions. GC–MS analysis of the irradiated mixtures showed the formation of substituted quinolines as the main products when 3,5-dimethyl-nitrobenzene and 3-nitrotoluene were employed as substrates, while the ethoxy-tetrahydroquinolines were obtained when 2- and 4-nitrotoluene were used [12].

Very recently, Selvam and Swaminathan have reported the synthesis of 2-methylquinolines (quinaldines) from nitroarenes (nitrobenzene, 3- and 4-nitrotoluene, 3,5-dimethylnitrobenzene, 4-methoxy-, 4-chloro-, and 4-fluoro-nitrobenzene) using Au@TiO$_2$ as the photocatalyst in water-free ethanol under continuous purging with molecular nitrogen. Under the experimental conditions of their work 99% of the nitroarene were converted within five hours of UV(A) irradiation. The yields of the 2-methylquinoline were usually 60 – 80%, while the nitroarenes having a halogen in the para-position were transformed to their respective 2-methylquinolines with much lower yields (<20%). Employing bare TiO$_2$ also resulted in much lower yields (40 – 70%) for the conversion of nitrobenzene, 3-nitrotoluene, and 4-methoxynitrobenzene to the corresponding quinaldine. The reaction of nitrobenzene in the presence of Au@TiO$_2$ was investigated in more detail. Increasing the concentration of water in the solvent from 0 to 4 vol-% resulted in a decrease of the yield of 2-methylquinoline from 75 to 6% while the conversion of the nitro-benzene was nearly not affected. GC-MS chromatography revealed the formation of nitroso-benzene, N-hydroxyaniline, aniline, and 2-methyl-N-phenyl-1,2,3,4-tetrahydroquinoline-4-amine as intermediates during the quinaldine formation and the formation of 2-methyl-1,2,3,4-tetra-hydroquinoline as a secondary product [13].

To obtain some information about some factors influencing the chemical structure of the products and to elucidate the underlying mechanisms of the C–N- and C–C-coupling reactions mentioned above, the reactions of 3,5-dimethylnitrobenzene, 2-, 3-, and 4-nitrotoluene in O$_2$-free alcoholic suspensions of TiO$_2$, Pt@TiO$_2$, Pd@TiO$_2$, and Au@TiO$_2$ have been investigated in this work.

EXPERIMENT

All reactions were performed in glass snap-cap bottles (borosilicate glass, diameter = 23 mm, height = 75 mm) at ambient temperature under continuous stirring. 25 mg of the photocata-lyst, either pure TiO$_2$ (Sachtleben Hombikat UV100) or metal / TiO$_2$ (25 mg), was suspended in an alcoholic solution (10 cm^3) containing the nitroaromatic compound (100 μmol, 3,5-dimethyl-nitrobenzene, 2-, 3-, 4-nitrotoluene) and the required amount of the co-catalyst p-toluene sulfonic acid (0 - 40 μmol). The samples were irradiated by a 500-W mercury medium-pressure lamp Heraeus TQ 718 Z4 (UV(A) light intensity = 15 W m^{-2}) for two hours after purging the suspen-sion with Argon for 30 minutes. Metal-loaded TiO$_2$ photocatalysts, i.e. Pt@TiO$_2$, Pd@TiO$_2$, and Au@TiO$_2$ were prepared by published photocatalytic deposition methods in aqueous suspensions of Sachtleben Hombikat UV100 [14]. GC–MS chromatography was used for qualitative and quantitative analyses after removing the semiconductor particles through filtration (0.45 μm).

RESULTS

The UV(A) light-induced conversion of 3,5-dimethylnitrobenzene **1a** (5-nitro-m-xylene), 2-nitrotoluene **1b**, 3-nitrotoluene **1c**, and 4-nitrotoluene **1d** in Argon-purged, O$_2$-free ethanolic sus-pensions of a photocatalyst yields a variety of products depending on the type of photocatalyst (bare TiO$_2$ or metal-loaded TiO$_2$) and the presence of an acidic co-catalyst. Scheme 2 depicts an overview over the intermediates and products while Table I summarizes the conversion of the nitroarene **1** and the selectivity for the most frequent reaction products detected by GC–MS chro-matography of the reaction mixture.

The UV(A) light-induced conversion of the nitroarenes **3a–d** in argon-purged ethanolic suspensions of TiO$_2$ in the presence of p-toluenesulfonic acid (TSA) as an acidic co-catalyst yielded ethanal as the primary oxidation product of ethanol, anilines **2a–d** as the reduction products of the nitroarene, imines (Schiff bases) **3a–d** as the products resulting from C–N-coupling, and the heterocyclic compounds **4a–d**, **5a–d** and **6a–d** in varying amounts as the products of subsequent C–C-coupling reactions. These reactions are not specific for ethanol / ethanal as revealed by the respective photocatalytic conversion of **1a** in suspensions of TiO$_2$ in other primary alcohols (vide Table II).

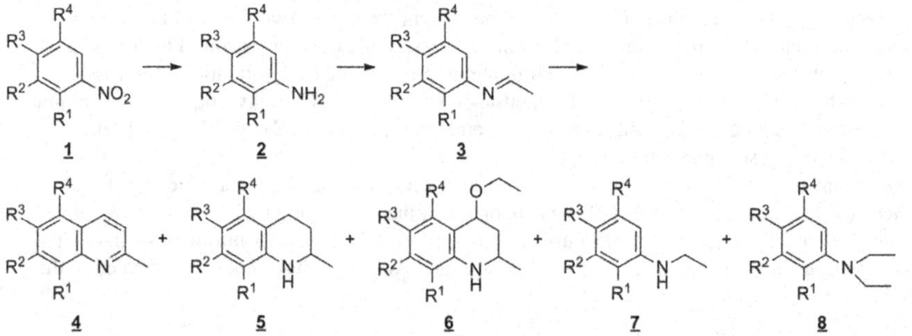

Scheme 2. The products of the UV(A) light-induced transformation of nitroarenes in O$_2$-free ethanolic suspensions of a TiO$_2$-based photocatalyst.

Table I. Products of the reaction of the nitroarenes **1** in ethanolic suspensions of TiO$_2$ and metal-loaded TiO$_2$.

Substituents				Cat.	Conv. /%	Selectivity / %						
R^1	R^2	R^3	R^4			**2**	**3**	**4**	**5**	**6**	**7**	**8**
a H	CH$_3$	H	CH$_3$	TiO$_2$, TSA	100			74.0	4.5			
				Pt@TiO$_2$	100			9.1			74.4	13.7
				Pd@TiO$_2$	41.1			28.8			65.1	6.1
				Au@TiO$_2$	8.5	79.1	20.9					
b CH$_3$	H	H	H	TiO$_2$, TSA	100	30.7	44.4		5.7	19.2		
				Pt@TiO$_2$	94.3	7.8	17.0	1.0	2.0	4.1	64.0	4.1
				Pd@TiO$_2$	40.2	9.0	13.8		2.0	1.0	73.8	0.5
				Au@TiO$_2$	20.3	45.9	54.1					
c H	CH$_3$	H	H	TiO$_2$, TSA	100			89.9		10.1		
				Pt@TiO$_2$	100			24.7			60.8	14.5
				Pd@TiO$_2$	13.1			17.9			82.1	
				Au@TiO$_2$	21.3		59.2		40.8			
c H	H	CH$_3$	H	TiO$_2$, TSA	100			19.9	80.1			
				Pt@TiO$_2$	100			3.5		2.4	21.1	72.9
				Pd@TiO$_2$	28.7			6.2			91.2	2.6
				Au@TiO$_2$	14.0	29.7	70.3					

Experimental conditions: 100 µmol **1**, 25 mg photocatalyst (bare or metal-loaded (0.5 wt-%) Sachtleben Hombikat UV 100 TiO$_2$) and 40 µmol p-toluenesulfonic acid if required in 10 ml alcohol; reaction time = 2 h, UV(A) light intensity = 15 W m^{-2}.

Table II. Products of the photocatalytic reactions of 3,5-dimethylnitrobenzene **1a** in suspensions of TiO_2 in different primary alcohols.

alcohol	main product	by-product
ethanol	**4a**	**5a**
1-propanol	**9**	
1-butanol	**10**	
3-methyl-1-butanol	**11**	

Experimental conditions: 100 µmol **1a**, 25 mg Sachtleben Hombikat UV 100 TiO_2, and 40 µmol p-toluenesulfonic acid in 10 ml alcohol; UV(A) light intensity = 15 W m^{-2}.

A plot of the concentrations of the aromatic compounds present in the reaction mixture versus irradiation time (not shown) evinces that the nitroarene **1** is rapidly reduced to the corresponding aniline **2**. Maximum concentrations of **2** are reached at the time when the concentration of the starting compound **1** approaches zero. At longer irradiation times the concentration of **2** decreases. The concentration of the heterocyclic compounds **4** – **6** increases with irradiation time, indicating that these compounds are formed in secondary reactions of the aniline **2**. To identify the photocatalytic step(s) during the formation of the quinolines **4** and related heterocyclic compounds from nitroarenes and primary alcohols, some experimental runs in the dark were performed with ethanolic solutions of 3,5-dimethylaniline **2a** and ethanal in access. The thermal reaction between **2a** and ethanal at ambient temperature in the dark without any added catalyst yielded 2,5,7-trimethylquinoline **4a** and 4-ethoxy-2,5,7-trimethyl-1,2,3,4-tetrahydroquinoline **6a**. In the presence of TiO_2 the same reaction gave only small amounts of **4a** but much higher amounts of **6a**. Addition of p-toluenesulfonic acid as an acidic catalyst to a mixture of **2a** and ethanal in ethanol increased the amount of **4a**. A small amount of 2,5,7-trimethyl-1,2,3,4-tetrahydroquinoline **5a** but no **6a** was detected by GC-MS chromatography. The same products were also found during the dark reaction between the aniline **2a** and ethanal in the presence of TiO_2 and p-toluene sulfonic acid. Most likely, traces of 4,5,7-trimethylquinoline were formed in these four experimental runs.

It is well known that noble metals deposited on the TiO_2 surface act as electron sinks for the photogenerated conduction band electrons thus facilitating the reduction of adsorbed nitroarenes [15]. Therefore, Pt, Pd, and Au were deposited on the Hombikat UV 100 TiO_2 surface by published methods [14] and the light-induced reactions of the nitroarenes **1a–1d** in ethanolic suspensions of these three photocatalysts were investigated. The main products of these reactions as identified by GC-MS chromatography and the selectivity of their formation are given in Table I. As can be seen from this table, Pt@TiO$_2$ favours the formation of the N-alkylanilines **7a–d**. After prolonged irradiation time the N-alkylanilines are alkylated to the N,N-dialkylanilines **8a–d**. Only small amounts of quinolines **4** were detected.

The formation of alkylated anilines is not a specific reaction of ethanol / ethanal. As was shown by the light-induced reaction of 4-nitrotoluene **1d** in suspensions of Pt@TiO$_2$ in different primary C2 – C4 alcohols the respective N-alkylated and N,N-dialkylated derivatives were formed (vide Table III).

Pd@TiO$_2$ seems to posses a much lower reactivity than Pt@TiO$_2$, but like the latter this photocatalyst favours the formation of N- alkylated anilines (vide Table I). Au@TiO$_2$ exhibits also a low reactivity for the reduction of nitroarenes. The main product detected after two hours of irradiation of an ethanolic suspension containing a nitroarene and Au@TiO$_2$ was usually the corresponding imine **3**.

From the time course of the light-induced reaction (data not shown) it becomes obvious that the N,N-dialkylated compounds **8** are secondary products formed by alkylation of the N-alkyl-arylamine **7**.

Table III. Products of the photocatalytic reactions of 4-nitrotoluene **1d** in suspensions of Pt@TiO$_2$ in different primary alcohols.

alcohol	products	
ethanol	**7d**	**8d**
1-propanol	**12**	**13**
1-butanol	**14**	**15**

Experimental conditions: 100 μmol **1a**, 25 mg platinized (0.5 wt-%) Sachtleben Hombikat UV 100 TiO$_2$ in 10 ml alcohol; UV(A) light intensity = 15 W m^{-2}.

DISCUSSION

The UV(A) light-induced photocatalytic transformation of nitroarenes in alcoholic (metalized) TiO$_2$ suspensions yields different C–N-coupling products depending on the respective reaction conditions. However, in all cases the only photocatalytic reaction steps are the 6-electron reduction of the nitroarene by conduction band electrons (reaction 1) and the concurrent oxidation of ethanol (reaction 2).

$$Ar–NO_2 + 6\,H^+ + 6\,TiO_2\{e^-\} \rightarrow Ar–NH_2 + 2\,H_2O + 6\,TiO_2 \tag{1}$$

$$3\,CH_3–CH_2OH + 3\,TiO_2\{h^+\} \rightarrow 3\,CH_3–CHO + 6\,H^+ + 3\,TiO_2\{e^-\} \tag{2}$$

124

While imines **3** were found as the sole C–N-coupling products during the photocatalytic reactions of nitrobenzene and alcohols in the presence of bare TiO_2 [9] it is supposed that the aniline, Ar–NH$_2$, reacts with ethanal, CH$_3$–CHO, yielding an imine in a subsequent thermal reaction (reaction 3).

$$Ar–NH_2 + CH_3–CHO \leftrightarrow Ar–N=CH–CH_3 + H_2O \tag{3}$$

While ethanal is formed in access (aldehyde : aniline = 3) the equilibrium is on the side of the products. It seems that the reactions 1 – 3 occur in suspensions containing bare TiO_2 as well as in suspensions of metalized TiO_2.

It was shown in an experimental run in the dark without any added catalyst that a thermal reaction at ambient temperature between an aniline and ethanal yielding a 2-methylquinoline **4** and a 4-ethoxy-2-methyl-1,2,3,4-tetrahydroquinoline **6** is possible when ethanal is present in high access. In the presence of TiO_2 the same reaction was found to yield only small amounts of **4** but much higher amounts of **6**, while the addition of an acidic co-catalyst (p-toluenesulfonic acid) increased the yield of **4**. In addition, small amounts of **5** but no **6** were detected in the presence of p-toluenesulfonic acid. These observations suggest that the subsequent transformation of the intermediate imine **3** to the heterocyclic compounds **4**, **5**, and **6** proceeds in a sequence of reactions including two C–C-coupling reactions and involving at least one acid-catalyzed reaction step to give high yields of **4**.

The acidity of some Ti–OH groups [16] is sufficient to act as the acidic catalyst but increasing the acidity of the suspension by adding an Arrhenius (Brönsted) acid (here: p-toluenesulfonic acid) favours the cyclization to the detected heterocyclic compounds. However, as has been shown by Selvam and Swaminathan [13] increasing the acidity of Ti–OH groups by depositing gold at the TiO_2 surfaces is also increasing the yield of quinolines.

On the other hand, metal deposits at the TiO_2 surface do not necessarily increase the yield of heterocyclic compounds as becomes obvious from the respective experiments employing platinum and palladium (vide Table I). In the presence of Pt@TiO_2 and Pd@TiO_2 the reaction pathway is changed favouring the formation of N-alkylarylamines **7** and – after prolonged reaction time – of N,N-dialkylarylamines **8**. It is well known, that noble metals like platinum and palladium at the TiO_2 surface act as sinks for the conduction band electrons thus hindering the electron-hole recombination and favouring the interfacial electron transfer to protons present in the surrounding solution [15]. Hence, the formation of the alkylated compounds **4** and **5** can simply be explained, assuming that electrons stored in the metal deposit (M{e$^-$}) react with protons yielding adsorbed hydrogen atoms (M–H$_{ads}$) (reaction 4) and in subsequent reactions either molecular hydrogen (reaction 5) is formed or a N-alkylarylamine by the reduction of the intermediate imine (reaction 6).

$$M\{e^-\} + H^+ \rightarrow M–H_{ads} \tag{4}$$

$$2\,M–H_{ads} \rightarrow 2\,M + H_2 \tag{5}$$

$$2\,M–H_{ads} + Ar–N=CH–CH_3 \rightarrow 2\,M + Ar–NH–CH_2–CH_3 \tag{6}$$

The N-alkylarylamine **7** can react in a susbsequent thermal reaction with another ethanal molecule yielding the N,N-dialkylated product **8**. Since bare TiO_2 is hardly able to reduce protons [15] the formation of N-alkylarylamines **7** is not observed in the absence of noble metal deposits. This finding is in agreement with experimental results published by Valenzuela and co-workers who have found imines **3** as the sole C–N-coupling product between photocatalytically formed aniline and ethanal in bare TiO_2 suspensions [9]. When $Au@TiO_2$ is used as the photocatalyst imines **3** are also found to be the main products under the experimental conditions of this work. Therefore, at the present time it is not possible to predict the further reaction(s) of the intermediate imines **3** after prolonged reaction time in the presence of $Au@TiO_2$. While gold is known to be only a poor hydrogenation catalyst it can be assumed that heterocyclic compounds will be the main products, a conjecture in accordance with the findings reported by Selvam and Swaminathan [13], who reported the highly selective formation of 2-methylquinolines **4** from nitroarenes in ethanolic suspensions containing $Au@TiO_2$.

CONCLUSIONS

Substituted quinolines and other heterocyclic compounds as well as N-alkyl- and N,N-di-alkylarylamines can be prepared from O_2-free alcoholic suspensions of nitroarenes by a TiO_2 mediated photocatalytic process. The composition of the product mixture strongly depends on the respective photocatalyst and the employed co-catalyst. An Arrhenius acid as co-catalyst is favouring the formation of substituted quinolines by a sequence of one C–N- and two C–C-coupling reactions while in presence of a platinum or palladium loaded TiO_2 the transformation of the nitroarene to an N-alkylarylamine is the main reaction pathway.

ACKNOWLEDGMENTS

A. Hakki thanks the Deutscher Akademischer Austauschdienst (DAAD), Germany, for granting him a PhD scholarship, and the Departement of Chemistry of the Damascus University, Syria, for granting him a leave of absence.

REFERENCES

1. B. Ohtani, B. Pal, and S. Ikeda, *Catal. Surv. Asia* **7**, 165-176 (2003).
2. G. Palmisano, V. Augugliaro, M. Pagliaro, and L. Palmisano, *Chem. Commun.* **2007**, 3425-3437.
3. Y. Shiraishi, and T. Hirai, *J. Photochem. Photobiol. C: Photochem. Rev.* **9**, 157-170 (2008).
4. G. Palmisano, E. García-López, G. Marcì, V. Loddo, S. Yurdakal, V. Augugliaro, and L. Palmisano, *Chem. Commun.* **46**, 7074-7089 (2010).
5. R. Galian, and J. Pérez-Prieto, *Energy Environ. Sci.* **3**, 1488-1498 (2010).
6. Mahdavi, T. C. Bruton, and Y. Li, *J. Org. Chem.* **58**, 744-746 (1993).
7. J. L. Ferry, and W. H. Glaze, *Langmuir* **14**, 3551-3555 (1998).
8. S. O. Flores, O. Rios-Bernij, M. A. Valenzuela, I. Córdova, R. Gómez, and R. Gutiérrez, *Top. Catal.* **44**, 507-511 (2007).

9. O. Rios-Bernÿ, S. O. Flores, I. Córdova, and M. A. Valenzuela, *Tetrahedron Lett.* **51**, 2730-2733 (2010).
10. A. Hakki, R. Dillert, and D. W. Bahnemann, presented at The European Materials Research Society Fall Meeting, 15-19 September, 2008, Warsaw, Poland, Book of Abstract, p. 103, and at the 24th International Conference on Photochemistry (ICP 2009), 19-24 July, 2009, Toledo, Spain (unpublished).
11. K. H. Park, H. S. Joo, K. I. Ahn, and K. Jun, *Tetrahedron Lett.* **36**, 5943-5946 (1995).
12. A. Hakki, R. Dillert, and D. W. Bahnemann, *Catal. Today* **144**, 154-159 (2009).
13. K. Selvam, and M. Swaminathan, *Catal. Commun.* **12**, 389-393 (2011).
14. B. Kraeutler, and A. J. Bard, *J. Am. Chem. Soc.* **100**, 4317-4318 (1978); J. Papp, H.-S. Shen, R. Kershaw, K. Dwight, and A. Wold, *Chem. Mater.* **5**, 284-288 (1993); G. Bamwenda, S. Tsubota, T. Nakamura, and M. Haruta, *J. Photochem. Photobiol. A: Chem.* **89**, 177-189 (1995).
15. D. Bahnemann, A. Henglein, and L. Spanhel, *Faraday Discuss.* **78**, 151-163 (1984); H. J. Zhang, G. H. Chen, and D. W. Bahnemann, *J. Mater. Chem.* **19**, 5089-5121 (2009).
16. H. P. Boehm, *Discuss. Faraday Soc.* **52**, 264-275 (1971).

Mater. Res. Soc. Symp. Proc. Vol. 1352 © 2011 Materials Research Society
DOI: 10.1557/opl.2011.1132

Photoactivity of Anatase–Rutile TiO₂ Nanocrystalline Mixtures Obtained by Heat Treatment of Titanium Peroxide Gel

Snejana Bakardjieva[1], Jan Subrt[1], Petra Pulisova[1], Monika Marikova[1] and Lorant Szatmary [1]

[1]Institute of Inorganic Chemistry ASCR v.v.i., CIT, 25068 Rez, Czech Republic

ABSTRACT

Nanosized titanium dioxide photocatalysts with anatase structure or mixture of anatase and rutile phases have been synthesized. Homogeneous precipitation of aqueous solutions containing TiOSO₄ and ammonia and following treatment with H₂O₂ was used to prepare porous yellowish (Ti-Per) gel. The gel was lyophilized for 48h and (Ti-Per)LYO-powder was obtained. Single phase anatase TiO₂ samples were prepared by heating of the (Ti-Per)LYO powder. The lamella-like particle morphology of TiO₂ samples determined by SEM were stable in air up to 950°C.The structure evolution during heating of the starting (Ti-Per)LYO powder was studied by DTA and XRD analyses in overall temperature range of phase transformation. The morphology and microstucture characteristics were also obtained by HRTEM and BET methods. The photocatalytic activity of the sample titania TI-LYO/950 heated to 950°C in air contained 78.4% anatase and 21.6% rutile was higher than standard Degussa P25. Titania sample TI-LYO/950 reveals the highest catalytic activity during the photocatalyzed degradation course of 4-chlorophenol in aqueous suspension under UV-irradiation.

INTRODUCTION

Photocatalysts are part of a group of advanced materials which have been intensively designated in the recent years. Among their broad applications in energy storage they are served as vital components in many commercial processes for removing environmental pollutants.
It is well known that photocatalytic reactions are occurred on the exterior boundary of samples that's why the surface properties of TiO₂ (such as pH, surface hydroxyl groups, crystalline phases or crystal defects) are regarded as the most important factors in determining the photocatalytic reaction kinetics and mechanisms [1]. In this study, we are using freeze drying method (lyophilization) leading to formation of TiO₂ materials with controlled morphology [2]. It is believe that anatase crystals prepared from peroxo-modified titanium dioxide gel have several exceptional properties. The gel is pure and contains no counter ions except those of water and peroxide, it is transparent and contains the Ti$^{(IV)}$O$^-_2$ radical in both aqueous and dried states [3-4].The formation mechanism of the Ti-peroxy gel is not yet fully elucidated, but some theory postulated of bi-nuclear peroxititanum existence.The aim of this paper is to modify the surface structure of anatase TiO₂ and find the optimal experimental conditions due to obtain the photocatalyst with high activity in degradation of 4-chlorophenol (4-CP) as a model reaction.

EXPERIMENT

The preparation of TiO_2 nanoparticles was described as follow: $TiOSO_4$ (4.8 g) was dissolved in 150 ml water and cooled under stirring untill floating ice mixture was generated. NH_4OH was added to the mixture and pH = 8 was reached. The solution was kept at 0°C for 1 hour due to good aging of the precipitate. After washing with water and filtering off hydroxen peroxide H_2O_2 (30 ml, 30 mass%) was added drop by drop to the precipitate. The final pH=2 was achieved and porous yellowish (Ti-Per) gel was formed. The gel was lyophilized for 48h and (Ti-Per)LYO-powder was obtained. As prepared product was heated over a wide temperature interval from 200-950°C in order to find the phase transition temperature for anatase-rutile TiO_2 modification. Heating was continued for 1 hour in a laboratory muffle furnace at a rate of 10°C min^{-1} and seven new samples TI-LYO/200 - TI-LYO/950 were isolated. Diffraction patterns were collected with diffractometer PANalytical X'Pert PRO equipped with conventional X-ray tube (Cu Kα radiation 40kV, 30 mA).Transmission electron micrographs were obtained by HRTEM JEOL JEM 3010 operated at an accelerated voltage of 300 kV (LaB_6 cathode). The specific surface area of the samples was determined by nitrogen adsorption-desorption isotherm at liquid nitrogen temperature using the Coulter SA 3100 instrument outgasing for 15 min at 120°C. Kinetics of decay of 0.1377 mM solution of 4-chlorophenol (4-CP) in 60 mL of aerated aqueous suspension of anatase and anatase–rutile photocatalysts (1 g L^{-1}) were monitored. 4-chlorophenol (99%) was purchased from Fluka and used without further purification. The laboratory irradiation experiments were performed in a self-constructed photoreactor and was described elsewhere [5]. The TiO_2 mixtures were sonicated for 20 min with a sonicator (230 W, 30 kHz) before use. The pH of the resulting suspension was taken as the initial value for neutral conditions and under the experiment was kept at 7.00. The kinetic analysis was understood in terms of modified (for solid–liquid reactions) Langmuir–Hinshewood kinetic treatment. This mechanism results in pseudo-first-order kinetics at the photostationary state.

DISCUSSION

Differential thermal analysis (DTA)

Figure 1 shows the DTA results confirmed exothermic peaks arose out in the sample (Ti-Per)LYO during the heating. Two peaks are appeared at 79°C and 239°C respectively and are related to the release of chemical and physical water. The second peak could be also distributed to the decomposition of NH_4^+ due to existing of $(NH_4)_2[TiO_2(SO_4)_2]$ complex. The peak at 371°C might be indicated the beginning of the crystallization process from (Ti-Per)LYO sample to the single crystal phase anatase. The broad exothermic peak at 807°C confirms the transformation phenomena from anatase to the stable rutile TiO_2 phase.

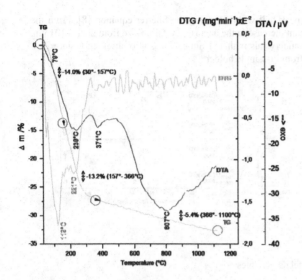

Figure 1. DTA analysis of (Ti-Per)LYO sample

X-ray diffraction

Powder X-ray diffraction (XRD) was used for crystal phase identification and the estimation of the anatase–rutile ratio.The crystallite size of phase TiO_2 mixture was also calculated. Our attention was focused on the inspection of the temperature interval from 300 to 950 °C where the anatase is formed, growth and intensive transforms to the stable rutile form. XRD pattern of sample (Ti-Per)LYO is shown on figure 2. XRD of the initial sample (trace a) indicated that there is no anatase TiO_2. This is the raw sample recovered at the end of 48 h of lyophilization. The characteristic parts of XRD patterns of the lyophilized-Ti product and its thermal treatment products at 200, 300, and 400°C are shown in figure 2 (traces b, c and d).The calcination at 200 and 300°C resulted in virtually no change in the intensity of all peaks, while the calcination at 400°C lead to the appearance of anatase (1 0 1) peak at $2\Theta = 25.4°$. The positions of all diffraction peaks and distribution of intensites correspond to ICDD PDF card 21-1272 [6].The XRD pattern of the powder (Ti-Per)LYO heated at 500, 650 and 800°C reveals that intensity of the anatase (1 0 1) significantly increase (figure 2,e,f,g traces). The rutile peak (1 1 0) at $2\Theta = 27.5°$ is registered at calcination of 950°C (figure 2, trace h).Sample (Ti-Per)LYO/950 contains 78.66% of anatase and 21.34% of rutile and the mixture looks like the standard product Degussa P-25 where the anatase:rutile relation is 4:1. The difference in the crystal structure in our titania sample with such anatase:rutile ratio is due to a thermal treatment at high temperature of the (Ti-Per)LYO and spontaneous transformation from anatase to rutile. Phase transformation anatase to rutile and calcination is usually accompanied with crystal growth [7]. The average size of crystallites of heated samples at 500,650,800 and

950C were calculated from the peak half-width B, using the Sherrer equation [8]. From the data obtained we can see significant increase in the anatase crystallite size from 26.1 – 101 nm. At temperature of 950°C rutile nanocrystals with 151 nm-size was also observed (figure 2,h). Diffraction lines noted as Pt are from Pt-sample holder.

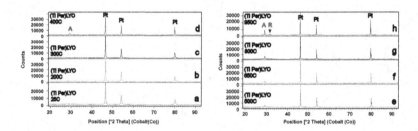

Figure 2. Powder X-ray analysis of TiO$_2$ samples

Particle morphology

SEM image of sample (Ti-Per)LYO heated at 950°C shows the fracture structure of lamellar material with small submicron islands situated on the surface (figure 3a). Aligned porous structure is also presented and according to reference [2] it may be produced during the lyophilization process. Thise morphology is originated as a result from the directional freezing process and it is stable during the heating. Figures 3b and 3c are presented the TEM low magnification of (Ti-Per)LYO sample heated at 950°C.The images reveal a unique "scorpio" or "antennae" type TiO$_2$ agglomerates consisting of uniform plate-like nanocrystals. Our results are in line with Ichinose et al. [9] finding that crystallization from titanic acid gel and hydrogen peroxide solution followed by lyofilization could be created regular 1D-sequence of ultrafine TiO$_2$ nanocrystals the size of most of which was below 20 nm.

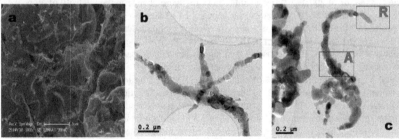

Figure 3. Particle morphology of (Ti-Per)LYO/950

Electron diffraction patterns in figure 4a illustrate fine grained anatase and rutile individual crystals with size about 20-25 nm. Figure 4b is shown ultrasmall square or hexagon shape nano-domains embedded in the surface of anatase or rutile plate crystals, respectively. HRTEM observation is confirmed that scorpio-like TiO_2 scaffold is made from both anatase tetragonal structure (SG I41/amd with lattice parameters a=3.7852Å and and c=9.5139 Å) and rutile tetragonal structure nanocrystals (SG P42/mnm with lattice parameters a=4.5933 Å and c=2.9592 Å) alternate each other (figure 4c).

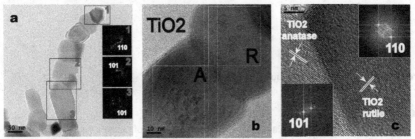

Figure 4. Electron diffraction and HRTEM images of (Ti-Per)LYO/950

Photocatalytic activity of TiO₂ in photodegradation of 4-CP

The course of 4-CP photocatalyzed decomposition is given on figure 4. We have found that the sample (Ti-Per)LYO/950 annealed at 950°C exhibited the best photocatalitic efficiency of the all tested TiO_2 samples. The photoactivity of this sample is 2.5 time higher than standard photocatalyst Degussa. The stoichiometric proportion between anatase:rutile there is the same such as in the standard 4:1. The particle size of both TiO_2 modification increased progressively: 101 nm for anatase and 151 nm for rutile nanocrystals. BET measurement is indicated surface area of 10.686 m^2/g. It is worth mentioned that sample calcinated at 800°C is also exhibited activity higher than the standard despite of its single phase anatase content.BET measurement is reaching surface area of 25.517 m^2/g. The photoactivity of sample (Ti-Per)LYO/650 consists of pure anatas nanocrystal with 33.7 nm size and BET surface of 31.907 m^2/g is comparable with the standard. These results assume that well crystallized plate-like TiO_2 nanocrystals with an average grain size between 30-100 nm and large surface area have generated interest because of the improvement of surface properties (no crystal defects) that expected to result in determining the photocatalytic reaction kinetics and mechanisms.

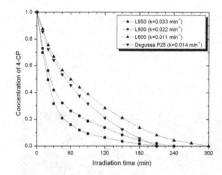

Figure 5. Photocatalytic decomposition of 4-CP

CONCLUSIONS

We have studied the formation of anatase or mixture of anatase-rutile TiO_2 from lyophilized titanium-peroxide gel. It has been demonstrated that freeze drying method leads to formation of 1D ordered TiO_2 nanocrystals with very high quality. The porous structure of TiO_2 materials is stable during the heating from 200-950°C. All samples with anatase structure are shown very good photoactivity in degradation of 4-CP. The best photocatalist is the sample similar to the standard Degussa P25 where the relation between anatase and rutile is 4:1.

ACKNOWLEDGMENTS

This research was supported in part by grants from the Ministry of Education, Youth and Sports of the Czech Republic IM4531477201 and LC523.We are expressed our special thanks to Ing. J.Bohacek and L. Volfova for providing the part of the synthesis and lyophilizations of the products.

REFERENCES

1. H. Park and W.Chio, J.Phys.Chem. **B109**, 11667-11674 (2005)
2. L.Qian, and H.Zhang, J.Chem. Technol.Biotechnol. **86**, 172-184 (2010)
3. P.Tengvall, T. P. Vikinge, I. Lundstrom and B. Liedberg, J Colloid Inter Sci.**160**,10-15 (1993)
4. P. Tengvall, B.Walivaara, J.Westerling and I.Lundstrom, J. Colloid Interface Sci.**143**, 589-591 (1991)
5. S. Bakardjieva, J. Subrt, V. Stengl, M.J. Dianez and M.J.Sayagues, Appl.Catal.B:Environm. **58**, 193-201 (2005)
6. JSPDS PDF release 2001, ICDD Mewtown Square, PA, USA
7. L.H.Edelson and A.M.Glaesr, J.Am.Ceram.Soc. **71**, 225-228 (1988)
8. R.A. Spurr and H.Mayers, Anal. Chem. **29**,760 (1957)
9. H. Ichinose, M. Terasaki and H.Katsuki, J Ceram Soc Japan. **104**, 715-718 (1996)

Mater. Res. Soc. Symp. Proc. Vol. 1352 © 2011 Materials Research Society
DOI: 10.1557/opl.2011.1011

High efficiency front-illuminated nanotube-based dye-sensitized solar cells

Kangle Li[1], Stefan Adams[1]
[1] Department of Materials Science and Engineering, National University of Singapore, 5
Engineering Drive 2, Singapore 117576

ABSTRACT

A highly reproducible two-step anodization method is reported to fabricate anatase TiO_2 nanotube layers. The nanotube membrane fabricated by this method is highly uniform and crack-free. Large area nanotube membranes can be transferred completely onto transparent FTO electrodes without the need for damaging ultrasonic agitation or acid treatment for application in front-illuminated nanotube-based dye-sensitized solar cells. A 16 μm thin front-illuminated nanotube-based dye-sensitized solar cell produced using this method reaches an efficiency of 6.3% under 1 sun illumination AM1.5.

INTRODUCTION

Dye-sensitized solar cells (DSC) are the most cost-effective photovoltaic device without requiring elaborate apparatus to manufacture [1]. As highly ordered, vertically oriented anatase TiO_2 nanotube arrays possess outstanding charge transport and carrier lifetime properties [2,3], their use as photo electrode for dye-sensitized solar cells [4,5] can help to boost the efficiency by improving the transport of the harvested charge carriers. Using nanotube arrays grown on Ti foils however requires a "back-illumination" design [6], where efficiency is limited by the need to illuminate the TiO_2 nanotube layer through the iodide-containing and thus absorbing electrolytes. Therefore, front-illumination DSC can greatly utilize the illumination without suffering the light reflection, scattering and absorption by the electrolyte. There are currently two established ways to fabricate front-illuminated nanotube-based dye-sensitized solar cell [7-9]. Sputtering of Ti metal on fluorine-doped tin oxide (FTO) followed by complete anodization of the metal layer increases the cost of fabrication and only limited thickness of Ti metal can be sputtered, typically less than 3 μm leading to TiO_2 nanotube membranes of suboptimal thickness < 5 μm [10]. The alternative method is to anodize a Ti metal foil and thereafter transfer the resulting titania nanotube membrane from the Ti foil onto FTO [11-13].

Previously, titania nanotube membranes were formed by anodizing thin Ti metal foils for several days until the metal foil is completely transformed [14]. For this method the nanotube membrane thickness is predetermined by the available Ti metal foils, so that the membranes are typically much thicker than what is expected to be suitable for front-illuminated DSC. When transferring nanotube membranes from a partially anodized metal foil, the main problem is to detach the membrane without damaging it. Of the two common techniques to detach nanotube membranes (ultrasonic agitation and acid treatment), ultrasonic agitation limits the reproducibility, as it typically results in cracking of the membranes rather than a detachment of an integer membrane. Cracking of the membrane was found to limit cell performance and a thick nanoparticle buffer layer (formed from the Ti isopropoxide) may cast doubt on the role of the nanotube layer [15,16]. On the other hand, acid treatment is highly sensitive to the thickness of membrane and may damage the nanotube microstructure. Without additional ultrasonication we also find that the acid treatment alone does not work reliably. If membranes do not quickly detach from the metal foil, the nanotube structure would be destroyed after long time immersion in the acid. In this paper, we demonstrate a new simple and highly reproducible method to fabricate front-illuminated nanotube-based DSC by detaching high quality nanotube membranes of tunable thickness in a multi-step anodization process, so that there is no need for ultrasonic agitation or long time acid treatment.

EXPERIMENT

Ti metal foil ($30 \times 15 \times 0.25$ mm 99.7%, Aldrich) and fluorine-doped tin oxide (FTO 15 Ω/sq., Dyesol) were cleaned in the sequence of 0.5% Decon 90, acetone, ethanol and deionized water 10 min each in an ultrasonic bath. 0.6wt% ammonium fluoride (Aldrich) in ethylene glycol (EG 99.8% anhydrous, Aldrich) aqueous solution was used for anodization (2% vol. H_2O). The cleaned and dried Ti foil was immersed in the electrolyte with a Pt foil (0.125 mm 99.99%, Aldrich) at a distance of 3 cm acting as the counter electrode. The first anodization was conducted under 50 V for 0.5 - 4 h with an initial voltage ramp rate of 0.05 V/s. 30 s ultrasonic agitation in a slurry of alumina powder [17] in DI water proved to be helpful in removing the debris on top of the nanotube layer. As-anodized foils were then dried at 200°C for 15 min, followed by 40 min annealing at 480°C to transform the initially amorphous TiO_2 into anatase, which is less soluble in the fluoride electrolyte during second anodization. The annealed foils were anodized again, now under 70 V for 20 min with a steeper initial voltage ramp rate of 1 V/s. This fast ramping promotes crack formation in the underlayer (and thereby the detachment of the membrane) while the slower ramp for the first anodization minimizes crack formation yielding more perfect nanotube membranes. After reaching constant voltage, the current initially decreased but after a few minutes the previously transparent membrane turned opaque white and the current started to increase again, indicating the formation of fresh Ti surface by a gradual detachment of the membrane. Once current density reached ~7.5 mA/cm^2, the complete nanotube membrane was detached from the Ti foil. The foil with the weakly adsorbed nanotube membrane was thereafter transferred to an isopropyl alcohol (IPA) containing Petri dish.

20 µl of a solution containing Titanium isopropoxide (97%, Aldrich) in IPA with triton x-100 (Aldrich) and acetic acid as additive in volume ratio 1:20:4:2 was spin-coated at 3000 rps on a 2×2 cm^2 FTO for 1 min. This layer helps to bond the nanotube membrane and inhibits recombination between FTO and electrolyte. The spin-coated FTO was annealed at 480°C for 30 min to form a 50 nm buffer layer of anatase. The annealed spin-coated FTO was then immersed in the same Petri dish. The nanotube membrane transfer from Ti metal foil onto FTO was conducted in IPA to minimize surface tension and damage to the membrane after exposure to air. By just tilting the metal foil, the titania nanotube membrane gently slides down onto the FTO. This avoids damages to the membrane by any mechanical action (e.g. by tweezers). The membrane-FTO stack was then removed from IPA and two drops of the above-mentioned isopropoxide IPA solution were used to fix the membrane on the FTO. TiO_2 adhesion layer was formed by infiltrating of titanium isopropoxide IPA solution or poly ethylene glycol solution (volume ratio 1:10) between anatase buffer layer and nanotube membrane. The nanotube on FTO was subsequently heated to 200°C for 20 min, and finally annealed at 480°C for 30 min. In this work, front-illuminated DSC devices with the structure FTO/nanotube/dye/electrolyte/Pt are abbreviated as FIF, and back-illuminated devices with the structure semitransparent Pt film /electrolyte/ dye/ nanotube/Ti foil are abbreviated as BIT. The number following this acronym specifies the duration of anodization in hours (e.g. FIF4 corresponds to front-illuminated cell with a nanotube membrane produced by 4 h anodization). The fabrication process of BIT cells follows the same procedure, except that second anodization and membrane transfer are skipped. DSC assembly and characterization details are described in our previous work [18].

RESULTS AND DISCUSSION

As described in the experimental section, nanotube membranes easily detach from Ti metal foil after second anodization. It may be expected that the compressive stress at the metal/film interface induced by the fast second anodization will be the main reason for the easy detachment. It may be noted that this mechanism differs from the O_2 gas bubble assisted delamination process recently reported by Wang et al. [19] that involves the breaking of the nanotubes. The free-flowing membrane shown in figure 1(a) and 1(b) are transparent and intact. Membrane transfer was achieved by simply letting the nanotube membrane in the IPA solution slide from the Ti metal foil onto FTO. After the FTO slide with the nanotube membrane has been dried, two drops of titanium isopropoxide solution were applied to the nanotube layer on FTO in order to attach the membrane. It was found that after annealing the membrane was uniformly and tightly attached onto FTO as displayed in figure 1(c). A major advantage when compared to alternative nanotube membrane transfer techniques [7, 9] is that by our method we could achieve a complete membrane transfer in more than 90% of the experiments. When using this highly repeatable method, the size of the detached membrane is only limited by the area of the Ti foil that is immersed in the anodization electrolyte (in the experiment 2×2 cm^2). It should be straightforward to scale up the method to a scale of interest for commercial application.

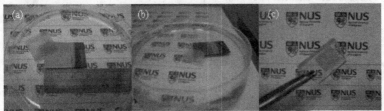

Figure 1. 16 µm thick nanotube membrane resulting from four hours anodization (a) and 4.9 µm thin nanotube membrane from 30 min anodization (b) flowing in IPA containing Petri dish after detachment from Ti foil; (c) membrane from Fig (a) attached on FTO after annealing. Membrane in (b) is more transparent than membrane in (a) as light is less scattered by thin nanotube membrane fabricated by half hour anodization.

Titania nanotubes with their well-connected straight pathways for transport of the harvested charge carriers are known to be outstanding in charge transport compared with nanoparticle structures, where charge transport has to follow tortuous random pathways. Figure 2 demonstrates that the morphology and thickness of nanotube after second anodization, transferring and fixing the membrane hardly differs from the morphology of the nanotube layer on the Ti foil [18], except for a small quantity of TiO$_2$ nanoparticles from the Ti isopropoxide solution used to attach the membrane on FTO. Though this solution is applied to the side of the membrane, it appears inevitable that it also reaches its top and decomposes to TiO$_2$ nanoparticles there. The fact that all nanotubes are closed at the bottom and some blocked by the accumulation of nanoparticles on the open top side (compare figures 2(a) and (b)) limit dye adsorption. As seen from the SEM cross section in figure 2(a), the 4.9 µm thick nanotube layer is firmly attached and vertically oriented on the FTO.

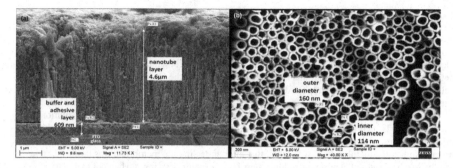

Figure 2. Side view (a) of well oriented nanotube arrays attached on FTO fabricated by half hour anodization yielding a thickness of 4.6 μm in line with the thickness measured by surface profilometer (4.9 μm); on average, FTO thickness: 400nm; TiO_2 buffer layer plus adhesive layers: 200 nm (of which 50 nm belong to the buffer layer as measured by surface profilometer); (b) top view of nanotube array prepared by half hour anodization on Ti metal foil with inner diameter as 114 nm and outer diameter as 160 nm.

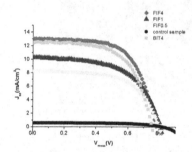

Figure 3. J-V characteristic of front-illuminated DSC based on nanotube membrane prepared by 4h (red), 1h (blue), 0.5 h (yellow) anodization and buffer layer only (black). For comparison, the J-V curve of a back-illuminated DSC is shown (green), for which the Ti foil has been anodized for 4 h as well. Active area of titania is 0.5 cm^2 in all cells.

Table I Photovoltaic performances under 1 sun illumination for nanotube-based DSC undergo several of anodization time.

	thickness	V_{oc} (V)	ff(%)	J_{sc} (mA/cm^2)	η(%)
FIF4	16 μm	0.78	63	13.0	6.3
BIT4	16 μm	0.75	60	12.4	5.7
FIF1	7.6 μm	0.80	56	10.2	4.6
FIF0.5	4.9 μm	0.80	60	8.3	3.9
Control sample	200 nm	0.81	60	0.55	0.27

J-V curves for the front-illuminated nanotube-based DSCs are shown in Figure 3. The resulting parameters characterizing the photovoltaic performance of front-illuminated nanotube-based DSCs with various nanotube lengths is summarized in table I and compared to the performance of a back-illuminated reference cell and a control sample without the nanotube layer. The nanotube length can be controlled via the anodization time [20], so that the method can produce nanotube membranes of tunable thickness. Particularly, membranes with thicknesses down to 4.9 μm could be fabricated by anodization for half an hour. Such thin nanotube membranes cannot be produced by the previously reported methods. This might be beneficial for boosting the efficiency of solid state front-illuminated nanotube DSCs, where recombination processes limit the effective diffusion length and thus shorter nanotubes become favorable. [21] In the case of the currently studied liquid electrolyte front-illuminated NT DSCs, efficiency keeps on increasing with nanotube length (for the limited range of anodization times from 0.5 to 4 hours studied so far), while open circuit voltage and fill factor are more or less the same.

Sample FIF4 prepared with the nanotube membrane obtained by 4 h anodization reaches an efficiency of 6.3% under AM 1.5 100 mW/cm^2 illumination with a promising open cell voltage and fill factor. The back-illuminated reference cell BIT4, which employs a nanotube layer anodized and annealed under the same conditions as FIF4 directly on the Ti metal foil, exhibits a lower efficiency of 5.7%. It has been tested separately by a control cell without nanotube membrane, but otherwise processed in the same way that the buffer layer formed by titanium isopropoxide only contributes 0.27% to this efficiency. So the difference between the front- and back-illuminated cells is larger than the effect of the buffer layer.

To gain a deeper insight into the origin of the performance difference of front- and backside illuminated DSCs (cf. table 1), a comparison of the monochromatic incident photo-to-electron conversion efficiency (IPCE) in figure 4 may be helpful. IPCE study of FIF4 and BIT4 demonstrates that the front illumination configuration permits the DSC cell to harvest energy over a wider range of wavelengths. This is essentially due to the absorption of wavelengths up to ca. 450 nm in the back-illuminated DSC by the iodine containing electrolyte [22]. We notice that the IPCE of FIF4 for the wavelengths from 500nm to 600nm is somewhat lower than to be expected, which might be due to light scattering in the buffer and adhesive layers and insufficient light absorption. Further optimization of the active nanotube layer, buffer layer and adhesive layer are in progress to enhance light harvesting efficiency to reach a plateau of ~80%.

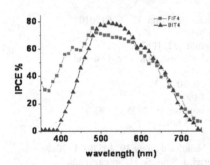

Figure 4. IPCE of front-illuminated FTO/NT/Pt DSC (Red) and back-illuminated Ti/NT/Pt DSC (Blue), each were prepared by 4 h anodization. Back-illumination suffers the light adsorption or reflection by electrolyte and Pt counter electrode, therefore photon conversion efficiency of front-illuminated cell outperforms back-illuminated cell particularly when the wavelength is from 350 nm to 450 nm. Higher photo-to-electron conversion efficiency of BIT4 than FIF4 from 500 nm to 600 nm is due to Ti reflection and light scattering by adhesive layer.

CONCLUSIONS

The nanotube membrane transfer method described in this work is highly repeatable without the need for using acid or ultrasonic agitation (which both tend to deteriorate the quality the nanotube morphology). Front-illuminated nanotube-based dye-sensitized solar cells with high light conversion efficiency (up to 6.3%) were fabricated. It is demonstrated that this light conversion efficiency nearly quantitatively originates from the nanotube layer. Electrochemical impedance spectroscopy was involved to reveal the electron transport characteristics in the nanotube layer: charge collection from the 4.9 – 16 μm thick nanotube membranes studied so far is highly efficient. Thus higher conversion efficiencies may be expected from thicker membranes. Further efficiency enhancements may be expected also from an optimization of the buffer and adhesion layers so as to reduce light reflection or absorption.

ACKNOWLEDGMENTS

Financial support to S. Adams by NUS AcRF 284-000-069-112 is gratefully acknowledged.

REFERENCES
1. B. O'Regan, M. Grätzel, Nature 353, 737 (1991).
2. K. Zhu, N.R. Neale, A. Miedaner and A.J. Frank, Nano Lett. 7, 69 (2007).
3. J.R. Jennings, A. Ghicov, L.M. Peter, P. Schmuki and A.B. Walker, J. Amer. Chem. Soc. 130, 13364 (2008).
4. C.M. Ruan, M. Paulose, O.K. Varghese, G.K. Mor and C.A. Grimes, J. Phys. Chem. B 109, 15754 (2005).
5. D. Kuang, J. Brillet, P. Chen, M. Takata, S. Uchida, H. Miura, K. Sumioka, S.M. Zakeeruddin and M. Grätzel, ACS nano. 2, 1113 (2008).
6. G.K. Mor, K. Shankar, M. Paulose, O.K. Varghese and C.A. Grimes, Nano Lett. 6, 215(2006).
7. J.H. Park, T.W. Lee and M.G. Kang, Chem. Commun. 25, 2867 (2008).
8. O.K. Varghese, M. Paulose and C.A. Grimes, Nat. Nanotechnol. 4, 592 (2009).
9. B.X. Lei, J.Y. Liao, R. Zhang, J. Wang, C.Y. Su and D.B. Kuang, J. Phys. Chem. C 114, 15228 (2010).
10. M. Paulose, L. Peng, K.C. Popat, O.K. Varghese, T.J. LaTempa, N.Z. Bao, T.A. Desai and C.A. Grimes, J. Membr. Sci. 319, 199 (2008).
11. Y. Jo, I. Jung, I. Lee, J. Choi and Y. Tak, Electrochem. Commun. 12, 616 (2010).
12. S.Q. Li, G.M Zhang, J. Ceram. Soc. Jpn. 118, 291 (2010).
13. J. Lin, J.F. Chen, and X.F. Chen, Electrochem. Commun. 12, 1062 (2010).
14. K. Shankar, G.K. Mor, H.E. Prakasam, S. Yoriya, M. Paulose, O.K. Varghese and C.A. Grimes, Nanotechnology 18, 065707 (2007).
15. C.J. Lin, W.Y. Yu and S.H. Chien, J. Mater. Chem. 20, 1073 (2010).
16. Q.W. Chen, D.S. Xu, J. Phys. Chem. C 113, 6310 (2009).
17. D. Kim, A. Ghicov and P. Schmuki, Electrochem. Commun. 10, 1835 (2008).
18. K.L. Li, Z.B. Xie and S. Adams, Z. Kristallogr. 225, 173 (2010).
19. D.A. Wang, L.F. Liu, Chem. Mater. 22, 6656 (2010).
20. Z.B. Xie, S. Adams, D. J. Blackwood and J. Wang, Nanotechnology 19, 405701-1 (2008).
21. U. Bach, D. Lupo, P. Comte, J.E. Moser, F. Weissortel, J. Salbeck, H. Spreitzer and M. Grätzel, Nature 395, 583 (1998).
22. S. Ito, S.M. Zakeeruddin, P. Comte, P. Liska, D.B. Kuang and M. Grätzel, Nat. Photonics 2, 693 (2008).

Mater. Res. Soc. Symp. Proc. Vol. 1352 © 2011 Materials Research Society
DOI: 10.1557/opl.2011.761

Photodeposition of Metal Sulfide Quantum Dots on Titanium(IV) Dioxide and its Applications

Hiroaki Tada
Department of Applied Chemistry, School of Science and Engineering, Kinki University, 3-4-1, Kowakae, Higashi-Osaka, Osaka 577-8502, Japan

ABSTRACT

In situ photodeposition techniques taking advantage of the TiO_2 photocatalysis have been developed for coupling metal sulfide quantum dots (QDs) and TiO_2 at a nonoscale. The coupled metal sulfide-TiO_2 systems possess the following characteristics: (I) a large amount of metal sulfides can be directly formed on TiO_2 during a fairly short period with excellent reproducibility, (II) the band energies of metal sulfides QDs are widely tunable by irradiation time, (III) metal sulfide QDs can be deposited on not only the external surfaces but also the inner ones of mesoporous TiO_2 nanocrystalline films without pore-blocking, (IV) the simple solution-based technique at low temperature enables the low-cost production, (V) this technique has a wide possibility for coupling TiO_2 and narrow gap metal sulfides. These unique features produce the excellent performances of the resulting heteronanojunction systems as the photoanodes for QD-sensitized solar cells.

INTRODUCTION

Narrow gap semiconductor represented by metal sulfide-TiO_2 coupling systems have attract much interest due to the possible applications to photocatalysts [1,2] and photovoltaics [3,4], where the common crucial process is the visible light-induced interfacial electron transfer (IET) from metal sulfide QDs to TiO_2. The self-assembled monolayer (SAM) technique using bifunctional coupling molecules is frequently used for the loading of metal sulfide QDs on the mesoporous TiO_2 nanocrystalline films (mp-TiO_2) [5,6], which is the key material of the dye- and QD-sensitized solar cells [7]. While this method allows us to precisely control the particle size, the loading level of QDs is limited below monolayer coverage, and thus the amount of light absorbed becomes low. Also, the direct contact of a large fraction of the TiO_2 surface with the electrolyte solution permits the back electron transfer from TiO_2 to the oxidant in the electrolyte solution. Further, the insulating molecules intervening between QD and TiO_2 at the junction could retard the IET [8,9]. Presently, the successive ionic layer adsorption and reaction (SILAR) technique [10] is believed to be the best way to prepare the metal sulfide QD-loaded TiO_2 films. On the other hand, since the discovery of the photodeposition of Pt on TiO_2 by Kraeutler and Bard [11], the photocatalytic synthesis has been developed for preparing TiO_2-based nanocomposites with metal complexes [12-14], polymers [15-17], and metal oxides [18-22]. Currently, it is being revealed to have a wide possibility of constructing the metal chalcogenide-TiO_2 coupled particle and film systems [23-32]. In the heterojunction systems, the efficient IET may be expected because the photoinduced redox reactions on the TiO_2 surface are taken advantage of for the preparation.

Here I summarizes the recent developments in the in situ photodeposition (PD) technique for preparing directly coupled metal sulfide-TiO_2 systems, and their application to the QDsensitize solar cells (QD-SSCs).

EXPERIMENT

A paste containing anatase TiO_2 particles with a mean size of 20 nm (PST-18NR, Nikki Syokubai Kasei) was coated on SnO_2-film coated glass substrates (12 Ω/□) by a squeegee method, and the sample was heated in air at 773 K to form mesoporous-TiO_2 films (mp-TiO_2/SnO_2). After mp-TiO_2/SnO_2 had been immersed into an ethanol solution (250 mL) containing S_8 (1.72×10^{-4} M) and $Cd(ClO_4)_2$ (1.38×10^{-2} M), and bubbled with argon for 0.5 h in the dark, irradiation was carried out for a given period with a high-pressure mercury lamp at 298 K; the light intensity integrated from 320 to 400 nm ($I_{320\text{-}400\,nm}$) was 3.7 mW cm^{-2}. After irradiation, the sample was recovered by centrifugation, and then washed with ethanol three times to be dried under vacuum. The resulting sample was treated with conc. HCl (10 mL), and the deposits were thoroughly dissolved into the solution by stirring for 1 h. The solution was diluted 5 times in volume with water, and then the Cd concentration was determined by inductively coupled plasma spectroscopy (ICPS-7500, Shimadzu). Cyclic voltammograms (CV) of TiO_2/SnO_2 were measured in ethanol solutions containing Cd^{2+} ions (1×10^{-3} M) or/and S_8 (1.25×10^{-4} M) and 0.1 M $NaClO_4$ supporting electrolyte under deaerated conditions using glassy carbon and Ag/AgCl as a counter electrode and a reference electrode, respectively. Photoelectrochemical cell (PEC) was designed using TiO_2/SnO_2 photoelectrode, a glassy carbon counter electrode, and a Ag/AgCl reference electrode. All photochronopotentiometry (PCP) measurements were carried out in a 0.1 M $NaClO_4$ electrolyte solution. After a constant potential had been reached by argon bubbling for 0.5 h in the dark, irradiation ($\lambda > 300$ nm, $I_{320\text{-}400\,nm} = 7$ mW cm^{-2}) was started by using a 300 W Xe lamp as a light source (Wacom HX-500). Electrochemical response with irradiation was followed for the PEC connected with a potentio/galvanostat (HZ-5000, Hokuto Denko). HRTEM observation and ED spectroscopic measurements were performed using a JEOL JEM-3000F and Gatan Imaging Filter at an applied voltage of 300 or 297 kV at an applied voltage of 300kV. The incident photon-to-current conversion efficiency (IPCE) was measured for sandwich-type photoelectrochemical solar cells (CdS/mp-TiO_2|SO_3^{2-}/S^{2-}|Au film) fabricated as follows. Au thin films with a thickness of ca. 100 nm were formed on 20 nm Cr-undercoated nonalkaline glass plates (NA35, Nippon Sheet Glass) by vacuum deposition. The cell gap was controlled at ca. 60 μm and the active area of the cell was 1.76 cm^2. The aqueous electrolyte solution of Na_2S (0.1 M), Na_2SO_3 (5.4×10^{-3} M) and $NaClO_4$ (0.1 M) was used after argon bubbling to remove the oxygen present in the solution. The short-circuit current (J_{sc}/A cm^{-2}) was measured using a potentio/galvanostat (HZ-5000, Hokuto Denko) as a function of excitation wavelength (λ/m), and the IPCE was calculated using eq (1).

$$\text{IPCE (\%)} = (J_{sc}\, N_A\, h\, c\, /\, I\, F\, \lambda) \times 100 \qquad (1)$$

where N_A is Avogadro constant, I (W cm^{-2}) is light intensity, F is Faraday constant, h is Planck constant, and c is speed of light.

Photocurrent-voltage (J-V) curves were measured under illumination by a solar simulator (PEC-L10, Peccell technologies, Inc.) at one sun (AM 1.5, 100 mW cm^{-2}) for the sandwich-type photoelectrochemical solar cells (photoanodes|3-methoxypropionitrile solution containing

0.1 M LiI, 0.05 M I$_2$, 0.6 M 1-propyl-2,3-dimethylimidazolium iodide, and 0.5 M 4-tert-butylpyridine |Pt). CdS(PD)/mp-TiO$_2$, CdS(SILAR)/mp-TiO$_2$, and CdS(SAM)/mp-TiO$_2$-L were used as the photoanodes. Prior to use, ZnS thin films were coated on the photoanodes by the following procedure [33]: the electrode was immersed in a solution of Zn(ClO$_4$)$_2$ (5.0 × 10^{-2} M) in water (20 mL) at room temperature for 1 min, and then the electrode was washed with water and dried in air. Subsequently, the electrode was immersed in a solution of Na$_2$S (5.0 × 10^{-2} M) in water (20 mL) at room temperature for 1 min, and then the electrode was washed with water and dried in air. The active area of the cell was 0.16 cm^2. The potentio/galvanostat (HZ-5000, Hokuto Denko) was used to record the J-V characteristics.

DISCUSSION

Photodeposition of metal sulfide QDs on TiO$_2$

As a typical example, the photodeposition of CdS QDs on TiO$_2$ is described [29]. UV-light irradiation to TiO$_2$ in an ethanol solution of Cd^{2+} ions and S$_8$ leads to the deposition of CdS to yield the directly coupled CdS-TiO$_2$ system at 298 K (CdS/TiO$_2$). A number of nanometer-sized particles are deposited on the TiO$_2$ surfaces in a highly dispersed state (Figure 1A). High resolution transmission electron microscopic (HRTEM) image shows two parallel lattice fringes, whose the nearest distances agree with the values for the (101) and (011) planes of hexagonal CdS (Figure 1B). Apparently, the interface exhibits good contact between TiO$_2$ and CdS with a fairly large contact area. In the energy dispersive X-ray spectrum (EDS) (Figure 1C), the signals of Cd and S are present besides those of Ti and O, while the Cu signals arise from the copper grid used for analysis.

UV-Vis absorption spectra of the samples prepared by changing irradiation time shows that new absorption due to the CdS interband transition appears in the visible region. With increasing irradiation time, the absorption intensifies, and the absorption edge significantly redshifts with (Figure 2A). The band gap (E_g) and mean particle size (d) of CdS were calculated as a function of irradiation time by the Tauc plot [34] and from the Brus equation (eq (2)) [35], respectively.

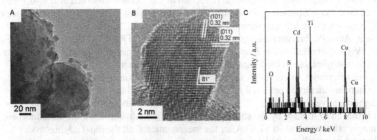

Figure 1. TEM (A), HRTEM (B) and energy dispersive X-ray (ED) spectrum (C) of a particulate sample obtained after 3 h irradiation: [Cd^{2+}]$_0$ =1.38 × 10^{-2} mol dm^{-3}, [S$_8$]$_0$ = 1.72 × 10^{-4} mol dm^{-3}. The figure is taken from ref. 29.

$$\Delta E_g = (\pi^2\hbar^2/2R^2)(m_e^{*-1} + m_h^{*-1}) - 1.8e^2/4\pi\varepsilon_0\varepsilon R \qquad (2)$$

where ΔE_g is a shift in E_g from the bulk E_g, R is the radius of CdS particle, m_e^* and m_h^* are the effective masses of electron and hole in CdS, respectively, ε_0 is vacuum permittivity, and ε is the relative permittivity of CdS.

The plots of the E_g and d (= 2R) as a function of irradiation time (Figure 2B) show that the E_g value of CdS increases with respect to the bulk value because of the quantum confinement in the rage of irradiation time below ca. 4 h. Both the loading amount and the E_g of CdS can be controlled by irradiation time in this process.

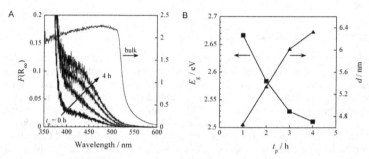

Figure 2. (A) UV-Vis absorption spectra of CdS/TiO$_2$ prepared by varying t_p: [Cd^{2+}]$_0$ = 1.38 × 10^{-2} M, [S$_8$]$_0$ = 1.72 × 10^{-4} M. (B) Plots of the E_g and d values as a function of irradiation time (t_p). $F(R_\infty)$ denotes the Kubelka-Munk function. The figure is taken from ref. 29.

The information about the photodeposition mechanism can be gained from (photo)electrochemical measurements [32]. CV curves of the mp-TiO$_2$ electrode were measured in deaerated aqueous electrolyte solutions (Figure 3A). In the presence of S$_8$ (a), a small reduction current flows in the range of the electrode potential (E) < -0.3 V. In the Cd(ClO$_4$)$_2$ solution (b), a much larger current due to the reduction of Cd^{2+} ions is observed at E < -0.1 V, accompanied by the corresponding oxidation with a current peak at -0.37 V. In the copresence of S$_8$ and Cd^{2+} ions (c), the reduction current starts to flow at $E \approx$ -0.1 V, whereas the oxidation peak is absent. Also, PCP measurements were performed for mp-TiO$_2$/FTO electrodes (Figure 3B). Upon UV-light irradiation, the E abruptly shifts towards the negative direction as much as -0.80 V, which is attributable to the Fermi energy upward shift resulting from the current doubling effect of ethanol. In curve (a), the first addition of Cd^{2+} ions shifts the E to -0.57 V close to the standard electrode potential of Cd^{2+}/Cd^0 (-0.60 V), and the subsequent S$_8$ addition causes a slight negative shift to -0.55 V. In curve (b), the first addition of S$_8$ hardly changes the E, while it is moved to -0.55 V by the subsequent Cd^{2+} addition. In curve (c), the simultaneous addition sharply shifts the E to -0.57 V. After the measurements, all the mp-TiO$_2$ electrodes changed from white to yellow, which indicates the CdS formation on the TiO$_2$ surface. The agreement of the E for the CdS formation with the standard electrode potential of Cd^{2+}/Cd^0 leads to the conclusion that the CdS photodeposition on TiO$_2$ progresses via the reduction of Cd^{2+} ion to Cd^0. Interestingly, the photodeposition of CdS on Au nanoparticle-loaded TiO$_2$ proceeds via

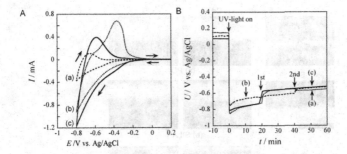

Figure 3. (A) Cyclic voltammograms of mp-TiO$_2$/FTO obtained for the first scan in the presence of S$_8$ (a), Cd^{2+} ions (b), and S$_8$ and Cd^{2+} ions (c): reference electrode = Ag/AgCl, [Cd^{2+}] = 1 × 10^{-3} M, [S$_8$]= 1.25 × 10^{-4} M, [NaClO$_4$] = 0.1 M. (B) PCP profiles of TiO$_2$/SnO$_2$ in a 0.1 M NaClO$_4$ aq: (a) Cd^{2+} ion addition at (1.38 × 10^{-2} M) at t_p = 20 min/S$_8$ (1.72 × 10^{-3} M) addition at t_p = 40 min, (b) S$_8$ addition (1.72 × 10^{-3} M) at t_p = 20 min/ Cd^{2+} ion addition (1.38 × 10^{-2} M) at t_p = 40 min, (c) simultaneous addition of Cd^{2+} ions and S$_8$ ([Cd^{2+}] = 1.38 × 10^{-2} M, [S$_8$]= 1.72 × 10^{-3} M) at t_p = 20 min. The figure is taken from ref. 29.

the preferential reduction of sulfur, yielding Au(core)-CdS(shell) nanoparticles on the TiO$_2$ surface [24].

Then, what kind of metal sulfides can be photodeposited on TiO$_2$? A useful information can be obtained from the standard Gibbs energy of metal sulfide formation ($\Delta_f G^0$) - metal redox potential (E^0 / V vs. SHE) diagram. The metal sulfides so far synthesized by the photodeposition technique (CdS, PbS, CuS, Ag$_2$S, MoS$_2$) are situated in the region of E^0 > -0.5 V and $\Delta_f G^0$ > -300 kJ mol^{-1}. Since the photodeposition of the metal sulfides is initiated from the reduction of the metal ions by the electrons photoexcited to the CB(TiO$_2$), the E^0 value must be more positive than the potential of the CB edge of TiO$_2$ (~ -0.5 V). On the other hand, the relatively high $\Delta_f G^0$ indicates that the metal sulfides with more covalent character than ionic character undergo the photodeposition. Fortunately, such metal sulfides have a narrow band gap, which is a prerequisite for the light absorbing material for the application to the solar energy conversion. In this manner, this technique has a wide possibility for coupling TiO$_2$ and narrow gap metal sulfides.

Characteristics of the photodeposition technique

The application of the photodeposition technique to mp-TiO$_2$ offers the additional unique and important characteristics [36]. Elemental depth profiles for CdS/mp-TiO$_2$ prepared by the photodeposition (PD) and SILAR methods indicate that both Cd and S are distributed towards the inner part of the mp-TiO$_2$ film (Figure 4). On the other hand, in the SAM sample, the CdS QD deposition on mp-TiO$_2$ occurs only at the thin upper part of the film. Pore size distribution in the mp-TiO$_2$ without CdS (Figure 5(a)) shows a peak pore size of *ca.* 20 nm. In the PD sample (b), the mesopores remain almost intact after the CdS deposition, although the peak diameter decreases approximately 35%. In contrast, in the SAM sample (c), the pore

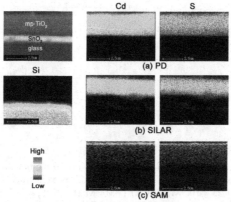

Figure 4. Elemental depth profiles by EPMA for CdS(PD)/mp-TiO$_2$ with t_p = 3 h (a), CdS(SILAR)/mp-TiO$_2$ with N = 7 (b), and CdS(SAM)/mp-TiO$_2$ with t_a = 24 h (c). This is taken from ref. 36

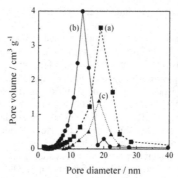

Figure 5. Pore size distribution plots determined by BJH analysis for the N$_2$ desorption branch of the isotherm from mp-TiO$_2$-S (a), CdS QD/mp-TiO$_2$-S (b) and CdS QD/MAA/mp-TiO$_2$-S (c).

size distribution hardly changes, and only pore volume significantly decreases. Further, the pristine mp-TiO$_2$ possessed a pore volume of 0.72 cm^3 g^{-1}, which was almost maintained after the CdS photodeposition. However, the value decreased to 0.20 and 0.25 cm^3 g^{-1} after depositing CdS by the SLAR and SAM methods, respectively.

The features of the photodeposition technique can be summarized as follows: (I) a large amount of metal sulfides can be directly formed on TiO$_2$ during a fairly short period with excellent reproducibility, (II) the band energies of metal sulfides QDs are widely tunable by irradiation time or surface modifier concentration, (III) metal sulfide QDs can be deposited on not only the external surfaces but also the inner ones of mp-TiO$_2$ without pore-blocking, (IV) the simple solution-based technique at low temperature enables the low-cost production, (V) this technique has a wide possibility for coupling TiO$_2$ and narrow gap metal sulfides suitable for the applications to the solar energy conversion.

Application to QD-SSC

In order to study the influence of the CdS-TiO$_2$ heterojunction state on the cell performance, IPCEs at λ_{ex} = 420 nm calculated from eq (1) were compared for the QD-SSCs using the CdS/mp-TiO$_2$ photoanodes prepared by the SAM, SILAR, and PD methods (Figure 6) [36]. The maximum IPCE is on the order of CdS(PD)/mp-TiO$_2$ (85%) > CdS(SILAR)/mp-TiO$_2$ (73%) >> CdS(SAM)/mp-TiO$_2$ (6.3%). The amount of CdS deposited by the PD and SILAR techniques are much larger than that by the SAM method. However, even at the same loading amount of CdS, the IPCE for CdS(SAM)/mp-TiO$_2$ is significantly smaller than those for CdS(PD)/mp-TiO$_2$ and CdS(SILAR)/mp-TiO$_2$. This difference points to the importance of the direct contact between metal sulfides and TiO$_2$ as well as the light absorption intensity for the increase in IPCE (feature I). The comparison between the direct coupling systems indicates that the IPCE of CdS(PD)/mp-TiO$_2$ exceeds that of CdS(SILAR)/mp-TiO$_2$ in spite of the smaller CdS loading amount. This superior performance of CdS(PD)/mp-TiO$_2$ over CdS(SILAR)/mp-TiO$_2$ can be attributed to the large contact area between CdS and the electrolyte solution because of its non-pore-blocking character (feature III). Also, there are optimum CdS loading amounts in the CdS(PD)/mp-TiO$_2$ and CdS(SILAR)/mp-TiO$_2$ systems. In these techniques, the loading amount and particle size of CdS concomitantly increase with an increase in irradiation time or SILAR cycle number. The higher CdS loading amount increases the light absorption. On the other hand, the increase in the CdS particle size decreases the driving force for the IET due to the lowering in the CB(CdS) (inset of Figure 6B), whereas the charge recombination can be suppressed [37]. In the CdSe/TiO$_2$ nanocoupling system, the electron-transfer rate constant has been shown to increase with decreasing CdSe particle size [38]. Thus, the balance between the light absorption intensity, the driving force for the IET, and the recombination process is considered as determining the optimum loading amount or particle size of CdS (feature II).

Power conversion efficiencies were evaluated for a sandwich-type cell consisting of the

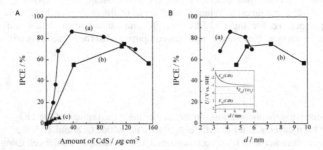

Figure 6. Plots of IPCE (λ_{ex} = 420 nm) vs. CdS loading amount for the cells using the following photoanodes: CdS(PD)/mp-TiO$_2$ (a), CdS(SILAR)/mp-TiO$_2$ (b), and CdS(SAM)/mp-TiO$_2$ (c). The figure is taken from ref. 36.

147

CdS/mp-TiO$_2$ photoanode | 0.1 M LiI + 0.05 M I$_2$ + 0.6 M 1-propyl-2,3-dimethylimidazolium iodide + 0.5 M 4-tert-butylpyridine (solvent = 3-methoxypropionitrile) | Au film counter electrode. Photocurrent-voltage curves for the cells using the photoanodes prepared by the PD, SILAR, and SAM techniques under illumination of one sun (AM 1.5, I = 100 mW cm^{-2}) were compared under each optimum condition. The open circuit voltage (V_{oc}), J_{sc}, fill factor (ff), and the power conversion efficiency (η) for the cells are summarized in Table 1. The η can be calculated by eq (3).

$$\eta\,(\%) = (J_{ph} \times V_{oc} \times ff\,/\,I) \times 100 \qquad (3)$$

A power conversion efficiency of 2.51% has been achieved for the sandwich-type solar cell using the CdS(PD)/mp-TiO$_2$ photoanode, whose value is much greater than those for the cells using the photoanodes prepared by the conventional SILAR and SAM techniques.

Table I. Cell performances obtained from the J_{ph}-V measurements. This table is taken from ref. 36.

Architecture	J_{sc} / mA cm^{-2}	V_{oc} / V	ff	η / %
CdS(PD) /mp-TiO$_2$ (t_p = 3 h)	6.53	0.69	0.56	2.51
CdS(SILAR)/mp-TiO$_2$ (N = 7)	2.73	0.70	0.64	1.21
CdS(SAM)/mp-TiO$_2$-L (t_a = 24 h)	0.49	0.64	0.46	0.14

The development of QD-SSCs are rapidly in progress, and the works hitherto performed have recently been reviewed [3,4]. Recent intensive researches towards the optimization of hole conductors, counter electrodes as well as the light absorbing QDs have achieved the QD-SSC efficiencies of 4-5% [39-41]. It is deserving special mention that a power conversion of 5.2% at 0.1 sun has been reported for a solid-state Sb$_2$S$_3$ QD-SSC using 2,2',7,7'-tetrakis(N,N-di-p-methoxyphenylamine)-9,9'-spirobi-fluorene as a hole conductor [42]. Further, a perspective for the QD-SSCs heightening absorber materials and their preparation methods, surface treatments, and naocomposite absorbers has recently been reported [43].

CONCLUSIONS

Simple low-temperature photodeposition techniques taking advantage of the TiO$_2$ photocatalysis have been developed for directly coupling metal sulfide QDs and TiO$_2$ at a nonoscale. The unique characteristics have been described. Owing to them, a solar cell using the QD-photodeposited mp-TiO$_2$ as a photoanode has provided the excellent performances. The present approach combined with photocatalysis and photoelectrochemistry should greatly contribute to the development of the solar energy conversion into electric and chemical energy.

ACKNOWLEDGMENTS

This work was supported by a Grant-in-Aid for Scientific Research (B) No. 20350097 from the Ministry of Education, Science, Sport, and Culture, Japan. The authors acknowledge T. Hattori and Y. Sumida (Nippon Shokubai Co.) for the EPMA measurements.

REFERENCES

1. H. Zhang, G. Chen, and D. W. Bahnemann, *J. Mater. Chem.* **19**, 5089 (2009).
2. G. Liu, L. Wang, H. G. Yang, H.-M. Cheng, and G. Q. Lu, *J. Mater. Chem.* **20**, 831 (2010).
3. S. Rühle, M. Shalom, and A. Zaban, *ChemPhysChem*, **11**, 2290 (2010).
4. P. V. Kamat, K. Tvrdy, D. R. Baker, and J. G. Radlich, *Chem. Rev.* **110**, 6664 (2010).
5. I. Robel, V. Subramanian, M. Kuno, and P. V. Kamat, *J. Am. Chem. Soc.* **128**, 2385 (2006).
6. Y. J. Shen, Y. L. Lee, and Y. M. Yang, *J. Phys. Chem. B* **110**, 9556 (2006).
7. B. O'Regan, and M. Grätzel, *Nature*, **353**, 737 (1991).
8. N. Mukaihata, H. Matsui, T. Kawahara, H. Fukui, and H. Tada, *J. Phys. Chem. C* **112**, 8702 (2008).
9. R. S. Dibbell, and D. F. Watson, *J. Phys. Chem. C* **113**, 3139 (2009).
10. R. Vogel, K. Pohl, and H. Weller, *Chem. Phys. Lett.* **174**, 241 (1990).
11. B. Kraeutler, and A. J. Bard, *J. Am. Chem. Soc.* **100**, 4317 (1978).
12. H. Tada, Y. Saito, and H. Kawahara, *J. Electrochem. Soc.* **138**, 140 (1991).
13. H. Tada, S. Tsuji, and S. Ito, *J. Colloid Interface Sci.* **239**, 196 (2001).
14. Y.-Y. Song, K. Zhang,and X.-H. Xia, *Appl. Phys. Lett.* **88**, 053112 (2006).
15. H. Tada, M. Hyodo, and H. Kawahara, *J. Phys. Chem.* **95**, 10185 (1991).
16. Y. Kim, J.-w. Lim, Y.-E. Sung, J.-b. Xia, N. Masaki, and S. Yanagida, *J. Photochem. Phtoobiol. A: Chem.* **204**, 110 (2009).
17. H.-Tang, J. Li, Y. Bie, L. Zhu, J. Zou, *J. Hazard. Mater.* **175**, 977 (2010).
18. Y. Matsumoto, M. Noguchi, and T. Matsunaga, *J. Phys. Chem. B* **103**, 7190 (1999).
19. K. Chiang, R. Amal, T. Tran, *Adv. Environ. Res.* **6**, 471 (2002).
20. E. Vigil, B. González, I. Zumeta, C. Domingo, X. Doménech, and J. A. Ayllón, *Thin Solid Films* **489**, 50 (2005).
21. M. Jin, X. Zhang, S. Nishimoto, Z. Liu, D. A. Tryk, A. V. Emeline, T. Murakami, and A. Fujishima, *J. Phys. Chem. C* **111**, 658 (2007).
22. N. Nishimura, J. Tanikawa, M. Fujii, T. Kawahara, J. Ino, T. Akita, T. Fujino, and H. Tada, *Chem. Commun.* 3564 (2008).
23. W.-Y. Lin, C. Wei, and K. Rajeshwar, *J. Electrochem. Soc.* **140**, 2477 (1993).
24. H. Tada, T. Mitsui, T. Kiyonaga, T. Akita, and K. Tanaka, *Nat. Mater.* **5**, 702 (2006).
25. Y. Tak, and K. Yong, *J. Phys. Chem. C* **112**, 74 (2008).
26. C. R. Chenthamarakshan, Y. Ming, and K. Rajeshwar, *Chem. Mater.* **12**, 3538 (2000).
27. S. Somasundaram, C. R. Chenthamarakshan, N. R. de Tacconi, Y. Ming, K. Rajeshwar, *Chem. Mater.* **16**, 3846 (2004).
28. V. N. H. Nguyen, R. Amal, and D. Beydoun, *J. Photochem. Photobiol. A: Chem.* **179**, 57 (2006).
29. M. Fujii, K. Nagasuna, M. Fujishima, T. Akita, and H. Tada, *J. Phys. Chem. C* **113**, 16711 (2009).

30. M. A. Zhukovskiy, A. L. Stroyuk, V. V. Shavalagin, N. S. Smirnova, O. S. Lytvyn, and A. M. Eremenko, *J. Photochem. Photobiol. A: Chem.* **203**, 137 (2009).
31. Y. Jin-nouchi, T. Akita, and H. Tada, *ChemPhysChem* **11**, 2349 (2010).
32. B. Ma, L. Wang, H. Dong, R. Gao, Y. Geng, Y. Zhu, and Y. Qiu, *Phys. Chem. Chem. Phys.* **13**, 2656 (2011).
33. S. Yang, C. Huang, J. Zhai, Z. Wang, and L. Jiang, *J. Mater. Chem.* **12**, 1459 (2002).
34. J. Tauc, R. Grigorovich, and A.Vancu, *Phys. Stat. Sol.* **15**, 627 (1966).
35. L. Brus, *J. Phys. Chem.* **90**, 2555 (1986).
36. Y. Jin-nouchi, S.-i. Naya, and H. Tada, *J. Phys. Chem. C* **114**, 16837 (2010).
37. Y. Tachibana, K. Umekita, Y. Otsuka, and S. Kuwabata, *J. Phys. Chem. C* **113**, 6852 (2009).
38. A. Kongkanand, K. Tvrdy, K. Takeuchi, M. Kuno, and P. V. Kamat, *J. Am. Chem. Soc.* **130**, 4007 (2008).
39. V. Gónzalez-Pedro, X. Xu, I. Mora-Seró, and J. Bisquert, *ACS Nano* DOI: 10.1021/nn101534y (2010).
40. S.-Q. Fan, B. Fang, J. H. Kim, J.-J. Kim, J.-S. Yu, and J. Ko, *Appl. Phys. Lett.* **96**, 063501 (2010).
41. J. A. Chang, J. H. Rhee, S. H. Im, Y. H. Lee, H.-J. Kim, S. I. Seok, M. K. Nazeeruddin, and M. Grätzel, *Nano Lett.* **10**, 2609 (2010).
42. S.-J. Moon, Y. Itzhaik, J.-H. Yum, S. M. Zakeeruddin, G. Hodes, and M. Grätzel, *J. Phys. Chem. Lett.* **1**, 1524 (2010).
43. I. Mora-Sero, and J. Bisquert, *J. Phys. Chem. Lett.* **1**, 3046 (2010).

Mater. Res. Soc. Symp. Proc. Vol. 1352 © 2011 Materials Research Society
DOI: 10.1557/opl.2011.875

Anodically Fabricated Sr-doped TiO2 Nanotube Arrays for Photoelectrochemical Water Splitting Applications

Hoda A. Hamedani,[1] Nageh K. Allam,[2] Hamid Garmestani [1], Mostafa A. El-Sayed[2]
[1]School of Materials Science and Engineering, Georgia Institute of Technology, Atlanta, GA 30332, U.S.A.
[2]Laser Dynamics Laboratory, School of Chemistry and Biochemistry, Georgia Institute of Technology, Atlanta, GA 30332, U.S.A.

ABSTRACT

The present work reports the synthesis of self-organized strontium-doped titania nanotubes arrays as a potential material for photocatalytic water splitting. Electrochemical anodization process was used to grow such material under various electrochemical conditions. The effect of dopant concentration on the morphology and photoelectrochemical properties of the material was investigated. The microstructure, morphology and composition of as-prepared and heat treated nanotubes were characterized by field emission scanning electron microscopy (FESEM), x-ray diffraction (XRD), transmission electron microscopy (TEM) and x-ray photoelectron spectroscopy (XPS). The results showed that increasing the dopant concentration up to its solubility limit results in higher photoelectrochemical activity. A preliminary proof of concept of the photocatalytic activity of the fabricated material was estimated in terms of the use of such material as a photoanode for photoelectrochemical water splitting.

INTRODUCTION

Photoelectrochemical water splitting has received great attention as an attractive way of hydrogen production from natural and renewable resources. Extensive research have focused on the development of semiconductor photocatalysts that have proper band gap (> 1.23 eV) and desirable relative energetic positions of the conduction and valence bands for water dissociation into hydrogen and oxygen. A wide range of transition metals such as Fe, Mo, Mg, Ag, Pt, Co, Cr, Mn and non-metal elements such as C, N, B, F have been used as dopants for the enhancement of the photoelectrochemical properties of TiO_2, which is the most promising photoelectrode material due to its low cost, nontoxicity, and photostability[1-5]. Recently, strontium titanate ($SrTiO_3$) has been intensively investigated as a photoanode for water splitting due to its high corrosion resistance, excellent photocatalytic activity, high stability, and non-toxicity. Fabrication of $SrTiO_3$ from amorphous TiO_2 nanotubes arrays have been performed using hydrothermal process that resulted in the formation of $TiO_2/SrTiO_3$ composites with improvement in the overall photoelectrochemical performance[6-8]. In this work, in-situ doping of TiO_2 nanotubes with strontium is investigated.

EXPERIMENT

The polished titanium foil (0.25 mm thick, Alfa Aesar) was rinsed in an ultrasonic bath of ethanol and cold D.I. water for 1 h and 15 min respectively. Pt mesh was used as the counter electrode in a two-electrode cell configuration with the Ti electrode immersed in the electrolyte. Samples were anodized in electrolyte containing 0.1M NH$_4$F and 0.02M, 0.04M, 0.06M anhydrous strontium hydroxide (Sr (OH)$_2$), 99% from Pfaltz and Bauer Inc. at 20 V for 3 h at 20 °C and the electrolyte pH was kept at 3. After anodization, the samples were rinsed in D.I. water and dried under stream of nitrogen followed by annealing at 450 °C for 24 h with heating/cooling rates of 1°C/min. The morphological characterization of the Sr-doped TiO$_2$ nanotubes was performed using a field emission scanning electron microscope (FESEM-Zeiss SEM Ultra60). The crystal structure of the phases was determined by glancing angle X-ray diffraction (GAXRD) using X'Pert PRO MRD diffractometer with Cu Kα radiation source. The surface properties and composition of the samples were analyzed by X-ray photoelectron spectroscopy using Thermo Scientific K-Alpha XPS with an Al anode.

Photoelectrochemical properties were investigated in 1.0 M KOH solution using a three-electrode configuration with Sr-doped nanotube arrays photoanodes, saturated Ag/AgCl as a reference electrode, and platinum foil as a counter electrode. A scanning potentiostat (CH Instruments, model CH 660D) was used to measure dark and illuminated currents at a scan rate of 10 mV/s. Sunlight was simulated with a 300 W xenon ozone-free lamp (Spectra Physics) and AM 1.5G filter at 100 mW/cm^2.

DISCUSSION

Figure 1 shows a typical FESEM top-view image of the fabricated Sr-doped TiO$_2$ nanotubes with diameters and lengths of 100±10 nm and 1.4 μm, respectively irrespective of the dopant concentration. However, it was observed that increasing the dopant concentration close to its solubility limit results in formation of precipitates not only in the electrolyte solution but also on the surface of the nanotubes.

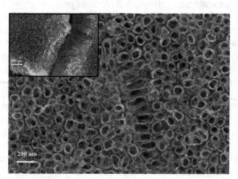

Figure 1. A top-view FESEM image and the inset show the cut-section view of the anodically fabricated Sr-doped TiO$_2$ nanotubes at 0.02M dopant concentration.

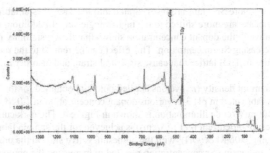

Figure 2. XPS survey spectra of Sr doped-TiO₂ nanotube with 0.02M dopant concentration.

Figure 2 shows the XPS survey spectra of the Sr-doped-TiO₂ nanotube with 0.02M dopant concentration confirming the presence of Sr, O and Ti in the sample. Figure 3 shows the GAXRD patterns of the annealed pure and Sr-doped TiO₂ nanotube arrays prepared in electrolyte of pH3. The samples show crystalline TiO₂ (anatase phase) after annealing and partial formation of SrTiO₃ in Sr-doped samples with preferred growth of (100) planes of SrTiO₃ corresponding to the peak indexed at 2θ=23° after the heat treatment. Such a preferential growth as well as the crystallinity of the doped samples is clearly observed in the TEM diffraction pattern which corresponds to the sample with 0.02M dopant concentration[9]. In addition, incorporation of the Sr^{+2} with larger ionic radius (1.32 Å) compared to Ti^{4+} (0.75Å) into the crystal lattice of TiO₂ resulted in peak shift in corresponding peaks of TiO₂.

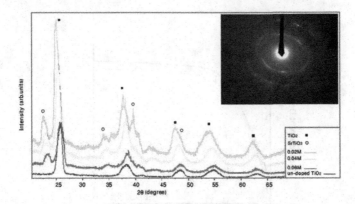

Figure 3. Glancing angle X-ray diffraction patterns of Sr-doped TiO₂ nanotube samples obtained at different Sr-dopant concentration in electrolyte of pH 3. The inset shows the TEM diffraction pattern of the sample with 0.02M dopant concentration.

The change in the intensity value of the diffraction peaks can be attributed to the variation of the crystallite size or change in concentration of the Sr ions. As the concentration of the dopant increases, more Sr^{+2} ions are incorporated into the TiO_2 crystal lattice and as a result the corresponding peaks are more shifted to the higher angle side. In addition, the increase in peak intensity relative to the dopant concentration shows that the crystallinity is slightly decreased with increasing Sr concentration. This effect can be related to the random substitution of the large Sr^{+2} ions in TiO_2 lattice that cause structural strain and disorder in TiO_2 crystalline lattice structure.

The photocurrent density (i_{ph}) versus applied potential obtained for the Sr-doped TiO_2 nanotube samples fabricated in pH 3 at various dopant concentration in 1.0 M KOH solution under AM1.5G (100 mW/cm^2) illumination is shown in Figure 4. The dark current in all cases is ~ 5 μA. The highest photocurrent (~ 1 mA/cm^2) is achieved for the sample fabricated in electrolyte containing 0.06 M Sr(OH)$_2$ with the highest negative shift in the onset potential (-0.803 V). Overall, increasing Sr concentration resulted in increasing the photocurrent; the trend observed in increasing the photocurrent with Sr concentration is analogous to the absorption capability of the material (not shown).

The corresponding light energy to chemical energy conversion (photoconversion) efficiency η was calculated using Eq. 1[10]:

$$\eta(\%) = [(\text{total power output-electrical power input})/\text{light power input}] \times 100$$
$$= j_p[(1.23 - |E_{measured} - E_{ocp}|)/I_0] \times 100 \qquad (1)$$

where j_p is the photocurrent density (mAcm^{-2}), $E_{measured}$ is the electrode potential (versus Ag/AgCl) of the working electrode at which the photocurrent was measured under illumination. E_{ocp} is the open circuit potential and I_0 is the intensity of the incident light (mWcm^{-2}).

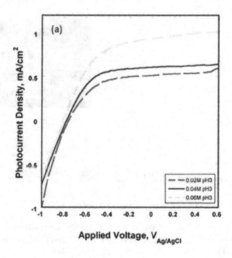

Figure 4. Variation of photocurrent obtained from the PEC measurements for S-doped TiO_2 nanotube electrodes with different Sr dopant concentration and electrolyte pH 3.

The highest photo-conversion efficiency of 0.69% was recorded for sample with the highest concentration of Sr dopant, i.e 0.06 M which is in agreement with the results of Sulaeman and co-workers where enhancement of i_{ph} activity under visible light irradiation in strontium titanate is observed for the material with high Sr/Ti atomic ratio[11]. According to Liu and co-workers[5], the highest peak energy conversion efficiencies reported to date have been 0.60% or less over the whole solar spectrum; thus, the obtained photoconversion efficiency of 0.69% for Sr-doped TiO_2 nanotubes in this work shows promising improvement towards a higher photoconversion efficiency.

CONCLUSIONS

In-situ fabrication of highly ordered Sr-doped TiO_2 nanotube arrays via electrochemical anodization technique for photoelectrochemical water splitting is reported. Nanotube arrays of Sr-doped TiO_2 have been fabricated and characterized at various concentrations of $Sr(OH)_2$. Nanostructural characterization of as-prepared Sr-doped TiO_2 nanotubes indicated that the quality of the nanotubes is dependent on the dopant concentration which is limited by the solubility limit of $Sr(OH)_2$ in the electrolyte. It is observed that Sr doping of TiO_2 nanotubes up to 0.06M can improve the photoconversion efficiency of the nanotubes up to 69%.

REFERENCES

1 N. K. Allam; M. A. El-Sayed, The Journal of Physical Chemistry C, **114**, 12024, (2010)
2 L. Deng; S. Wang; D. Liu; B. Zhu; W. Huang; S. Wu; S. Zhang, Catalysis Letters, **129**, 513, (2009)
3 P. Hartmann; D.-K. Lee; B. M. Smarsly; J. Janek, ACS Nano, **4**, 3147, (2010)
4 L. Jie; Y. Xinyuan; L. Deliang, Advanced Materials Research, **113-114**, 1945, (2010)
5 M. Liu; N. d. L. Snapp; H. Park, Chem. Sci., **2**, 80, (2010)
6 K. Shankar; J. I. Basham; N. K. Allam; O. K. Varghese; G. K. Mor; X. Feng; M. Paulose; J. A. Seabold; K.-S. Choi; C. A. Grimes, The Journal of Physical Chemistry C, **113**, 6327, (2009)
7 M. Ueda; S. Otsuka-Yao-Matsuo, Science and Technology of Advanced Materials, **5**, 187, (2003)
8 J. Zhang; C. Tang; J. H. Bang, Electrochemistry Communications, **12**, 1124, (2010)
9 J. Zhang; J. H. Bang; C. Tang; P. V. Kamat, ACS Nano, **4**, 387, (2010)
10 L. Jie; Y. Xinyuan; L. Deliang, Advanced Materials Research, **113-114**, 1945, (2010)
11 U. Sulaeman; S. Yin; T. Sato, APPLIED PHYSICS LETTERS, **97**, 103102, (2010)

AUTHOR INDEX

SUBJECT INDEX

Printed in the United States
by Baker & Taylor Publisher Services